Mario Wolter

Charge Transport in DNA – Insights from Simulations

Charge Transport in DNA –
Insights from Simulations

by
Mario Wolter

Dissertation, Karlsruher Institut für Technologie (KIT)
Fakultät für Chemie und Biowissenschaften
Tag der mündlichen Prüfung: 19. Juli 2013
Referenten: Prof. Dr. Marcus Elstner, Prof. Dr. Wolfgang Wenzel

Impressum

 Scientific
Publishing

Karlsruher Institut für Technologie (KIT)
KIT Scientific Publishing
Straße am Forum 2
D-76131 Karlsruhe

KIT Scientific Publishing is a registered trademark of Karlsruhe
Institute of Technology. Reprint using the book cover is not allowed.

www.ksp.kit.edu

Print on Demand 2013

ISBN 978-3-7315-0082-7

Charge Transport in DNA -
Insights from Simulations

Zur Erlangung des akademischen Grades eines

DOKTORS DER NATURWISSENSCHAFTEN

(Dr. rer. nat.)

von der Fakultät für Chemie und Biowissenschaften

Karlsruher Institut für Technologie (KIT) - Universitätsbereich

genehmigte

DISSERTATION

von

Mario Wolter

aus

Blankenburg (Harz)

Dekan: Prof. Dr. Martin Bastmeyer
Referent: Prof. Dr. Marcus Elstner
Korreferent: Prof. Dr. Wolfgang Wenzel
Tag der mündlichen Prüfung: 19.07.2013

Table of Contents

A	Adenine
Å	Angström (= 10^{-10}m)
AMBER	Assisted Model Building with Energy Refinement
a.u.	Atomic Units
C	Cytosine
CT	Charge Transport/Transfer
DFT	Density Functional Theory
DFTB	Density Functional Tight-Binding
DNA	Deoxyribonucleic Acid
dsDNA	Double Stranded DNA
EC	Electronic Coupling
ESP	Electrostatic Potential
G	Guanine
GROMACS	Groningen Machine for Chemical Simulations
HOMO	Highest Occupied Molecular Orbital
IP	Ionization Potential
LINCS	Linear Constraint Solver
MD	Molecular Dynamics
MF	Mean Field (Ehrenfest)
MM	Molecular Mechanics
QC	Quantum Chemistry
QM	Quantum Mechanics
SCC	Self Consistent Charge
SH	Surface Hopping
T	Thymine

Zusammenfassung

In dieser Arbeit werden die Eigenschaften von Ladungstransport und Ladungstransfer (LT) in Desoxyribonukleinsäure (DNA) untersucht. Besonderes Augenmerk wird hierbei auf die Struktur und die Ladungstransporteigenschaften der DNA unter experimentellen Bedingungen gelegt. Eine Mehrzahl der Experimente wird unter Bedingungen durchgeführt, die durch bisherige theoretische Untersuchungen nur unzureichend beschrieben werden.

Ein gemeinsames Merkmal von Ladungstransportexperimenten ist, die DNA zwischen zwei Elektroden mit variablem Abstand zu befestigen. Außerdem werden Einzelmolekülexperimente in sog. "Molecular Junctions" durchgeführt, hierbei wird das Lösungsmittel zum Teil entfernt.

Um die Ergebnisse der Experimente unter diesen Bedingungen erklären zu können, müssen einige wichtige Fragen beantwortet werden. Wieviel Wasser bleibt fest an die DNA gebunden und wie wird die Struktur beeinflusst? Wie verhält sich die DNA Struktur unter mechanischer Belastung? Was ist der Einfluss dieser strukturellen Veränderungen auf den LT? Warum hat die Struktur überhaupt einen Einfluss?

Alle diese Fragen werden im Laufe dieser Arbeit beantwortet. Um in der Lage zu sein, sie zu beantworten, wird dabei ein Simulations Setup erarbeitet, welches in der Lage ist den LT in DNA unter experimentellen Bedingungen zu beschreiben.

Auch wenn das verwendete Multiskalenmodell für die Beschreibung des LT eines der schnellsten verfügbaren ist, werden immernoch große Mengen von Rechenleistung benötigt. Um den LT in DNA Sequenzen aus hunderten von Nukleobasen über Zeitskalen größer als Nanosekunden zu simulieren, sind erhebliche Verbesserungen im Rechenaufwand erforderlich. Im letzten Teil dieser Arbeit wird daher ein möglicher Ansatz entwickelt - Ein parametrisiertes Modell für die Simulation des LT in DNA. Dieses Modell wird nicht nur in der Lage sein den LT richtig zu beschreiben, es wird außerdem Berechnungen an sehr langen DNA Sequenzen, bis zu tausenden von Nukleobasen, über ausgedehnte Zeiträume durchführen können.

Summary

In this work, the charge transport and charge transfer (CT) capabilities of deoxyribonucleic acid (DNA) are investigated. Special attention is drawn to the DNA structure and the CT under conditions resembling the experimental setup. The experimental investigations are conducted under conditions not represented accurately in most theoretical approaches until now.

A common feature of charge transport experiments is to attach the DNA strand to electrodes with varying distances. Moreover, single-molecule experiments are conducted in molecular junctions, where the solvent is removed partially with a stream of gas.

To explain the outcome of the experiments under such conditions several questions have to be answered. What is the exact amount of water preserved at the DNA strand and how does it affect the structure? How does the DNA structure accommodate for external mechanical stress? What influence have these structural changes on the charge transport/transfer? Why has it an influence at all?

All these questions will be answered in the course of this work. To be able to answer these questions a simulation setup is introduced, which is able to describe the experimental conditions in combination with the calculation of CT in DNA.

Although the used multi-scale framework for the computation of charge transfer in DNA is one of the fastest available, it still needs large amounts of computational time. To simulate charge transfer processes in DNA sequences of hundreds of nucleobases on time scales larger than nanoseconds, dramatic enhancements in computational efficiency have to be made. In the final part of this work, one possible approach to this is developed - a parametrized model to simulate the CT in DNA. Not only will this model be able to describe the CT in DNA accurately, it will make investigation of very long DNA sequences up to thousands of base-pairs over extensive periods of time possible.

CHAPTER 1

Introduction

The investigation of charge transfer processes constitutes an interface between physics, chemistry and biology. Chemical reactions are often accompanied by charge transfer processes not only in the laboratory, but also in biological processes. Here, DNA occupies an outstanding position in the researchers community. Its natural feature to support charge transfer is seen in processes like radiative damage and repair of DNA strands in living organisms. In the last years, the evolving field of molecular electronics has brought DNA into the focus as a potential element for electronic circuits at the molecular scale. Special features like self-assembly and self-recognition are a unique characteristic of DNA oligomers.

The conductivity of DNA has been studied intensively in the recent years.[1, 2] Yet, there are still features of charge transport (CT) that have not been resolved unambiguously yet. The research of CT in nucleic acids started with measurements performed on long DNA strands of several thousand base pairs, in the form of 'ropes'[3] or on supercoiled DNA.[4] In these early times, diametrically different conductivity of DNA under different conditions was reported, ranging from superconductivity[5] to insulation.[6]

The pioneering study of CT in individual DNA oligomers appeared in 2000,[7] and several further single-molecule experiments have been reported since then.[8–13] All of these studies are in a qualitative agreement upon the semi-conducting character of DNA under ambient conditions. The dependence of DNA conductance on the distance of the electrodes was proposed by Tao et al.,[8] however in a macroscopic fashion. Stepwise decrease of conductance was reported, as the individual DNA molecules between the electrodes were breaking.

The fundamental features of CT in DNA have been described as well. First, the exponential dependence of CT efficiency on the length of the DNA oligomers. Second, the role of experimental setup that is both crucial and challenging to characterize. In conclusion, carefully prepared DNA devices exhibited measurable conductance, and this fact together with the practicable synthesis of even very long DNA molecules were the initial impulses for the contemplations on the potential use of DNA-based devices in nano-electronics.[14, 15]

Albeit, the structure of the DNA species in these experiments could not be resolved unambiguously. While it was concluded that the structure played a key role in determining the conductance, it could only be guessed on the basis of observed electric properties what the structural contours were.

All mentioned experimental studies share the general features of the experimental setup. A DNA oligomer of several tens of base pairs is linked to metallic electrodes via a thioalkyl group attached usually to the 3'-end of each of the DNA strands. The distance of electrodes is or can be varied in the experiments, stretching the DNA molecule, and the CT efficiency is affected by the altered DNA structure. This fact directs the focus of the intended research to a considerable extent to the mechanical properties of the DNA species under the given conditions. The potential ability of DNA to stretch in biological processes was mentioned already in 1953,[16] in an article that directly followed the proposal of DNA structure by Watson & Crick.[17]

The response of dsDNA structure to stretching stress has been studied extensively since the 1990s. First considerations on entropic elasticity of DNA [18] were followed by studies on the DNA 'overstretching',[19, 20] and early modeling studies revealed that dsDNA would deform differently if the strands are pulled in different ways.[21, 22] Also, a structure was proposed that DNA would assume upon stretching of the 3'-ends of each strand – so-called S-DNA with maintained interstrand base pairing but unwound, in a sort of a ladder structure. On the other hand, when dsDNA was stretched and supercoiled, another new structure was observed.[23]

The state of knowledge of DNA stretching in 2000 was summarized in two reviews. [24, 25] At that time, a dispute on the nature of the overstretched state of dsDNA started, and it is in fact still going on. In contrast to the proposal of S-DNA, the overstretching profile of DNA was attributed to force-induced dehybridization based on thermodynamic observations.[26] Since then, reports have been constantly appearing against,[27] as well as in favor of the thermally induced melting.[28] Importantly, the nature of the overstretched state of DNA probably depends on the rate of stretching. A dsDNA oligomer was reported to dehybridize under a certain critical stretching rate, whereas a transition to S-DNA was observed above this threshold.[29] Further, DNA is observed more stable with 3′–3′ pulling than with 5′–5′, and it was pointed out that the relevant structures are different.[30] Two distinct overstretched DNA states were shown yet again to occur,[31] depending on conditions like DNA sequence, salt concentration and temperature. An excellent discussion of the state of the art on the DNA overstretching and of the current knowledge is given in Ref. [32].

As for the molecular modeling, quite a few studies on the overstretching transition of dsDNA appeared.[33–37] They all deal with dsDNA oligomers of up to 30 base pairs being stretched in different ways, pulling either the 5′ ends, the 3′ ends or eventually all ends of the double strand. Inevitably, the still substantial computational cost of atomistic molecular dynamics (MD) simulation makes the time scales probed in the experiments inaccessible. The previously reported dependence of the conformation of overstretched DNA on the loading rate is thus an issue that has to be taken into account when interpreting the outcome of MD simulations.

At this point, the dependence of CT in DNA on the conformation of the molecule can be investigated with computational methods conveniently. Electronic structure of overstretched DNA′ was the topic of an early study[38] dealing with a single sequence and static structures. Surprising response of CT efficiency on the DNA conformation was found,[39] however, with a simplified stretching/twisting reaction coordinate. Indeed, it has long been known that the dynamic character of molecular structure has to be taken into account instead of static structures.[40] Especially for more complex molecules like DNA, the structural dynamics and the molecular environment play decisive roles,[41] bringing these effects into the focus of thorough theoretical investigation.[42–46] Thus, the computational approach has to involve a plausible description of DNA conformation and its fluctuation under the conditions of applied mechanical stress, and this can be accomplished with classical MD simulation.

The dependence of CT on the (stretched) state of the DNA oligomer is of crucial importance for the construction of DNA-based nano-electronic devices. Here, a weak dependence of CT on the end-to-end distance within a certain interval is required.

The aim of this work is to determine how the efficiency of hole transport in DNA is affected by moderate mechanical pulling of the DNA molecule. Therefore, a microscopic interpretation will be sought for the outcome of DNA conductivity experiments, as the calculations are designed to resemble the usual experimental setup. Based on the results, suggestions will be made regarding the choice of nucleobase sequences in the design of DNA-based nano-electronic devices.

Common to the setup of many DNA conductivity studies [8, 9, 11, 12] is the fact that the aqueous solvent is removed so that the solvent conductivity does not bias the measurement. The amount of residual solvent attached to the DNA and its effect on the structural and electronic properties of DNA, however, remain unresolved up to date.

The dependence of DNA structure on humidity has been studied intensively since the 1950's.[47] However largely on macroscopic DNA fibers rather than in a single-molecule fashion with atomic resolution. The same is true about the effect of ion/salt concentration. Generally observed is the tendency of DNA to assume the A-like conformation under low humidity, contrary to the B-like one for 'wet' DNA. The mentioned conductivity experiments are an application where the effect of varied amount of water bound to a single DNA molecule on its structure and dynamics will need a more detailed investigation.

The solvent that the DNA assay had been prepared or kept in is removed to the largest possible extent, in order not to measure the conductivity of the buffer. Finally, the distance between the electrodes and thus the length of the DNA double strand may be varied. With such setup, the mentioned workers obtained valuable data on the electric properties of studied DNA species.

Although MD simulation seems to be a particularly useful tool to investigate DNA under the conditions of conductivity experiments, there has not been much work of this kind done. While this is by no means surprising if one considers the little biological relevance of such molecular systems, the recent interest in the electric properties of microhydrated DNA brings on a natural question – what is its structure? The only studies on a DNA species in a cluster with water molecules were performed by Mazur, who concentrated on another aspect though. A dsDNA in a quite large cluster of several thousand water molecules was simulated to show that the structure deviates little from the experimental one as well as from that observed in a simulation under periodic boundary conditions with a necessarily larger number of water molecules.[48] Also, reversible B–A conformational transitions were observed, depending of the (varied) amount of water in the cluster.[49] This amount was probably still too large compared with that bound to DNA in the single-molecule conductivity experiments.

However, the considerable experimental and theoretical research interest makes double-stranded DNA an archetypical system to study the dependence of CT efficiency on the molecular environment, especially the degree of hydration. Therefore, the first aim of this work is to describe the structure and mechanical stability of several short double-stranded DNA oligomers. Later on, the hole transport in double-stranded DNA oligonucleotides is studied in aqueous solution as well as under microhydration conditions. Possible consequences for CT in general will be discussed.

CHAPTER 2

Theoretical Background

The aim of theoretical or computational chemistry is to describe molecules and their properties as accurately and efficiently as possible. Unfortunately, with current computational power not both are possible at the same time for large molecules, e.g. biomolecules. Therefore, special methods for solving more or less specific tasks were developed. The methods at hand for description of the dynamics of large biomolecules are the molecular mechanics (MM), shortly described in chapter 2.1, and the molecular dynamics (MD), described in chapter 2.2. As this work does not only focus on the mere structures of biomolecules more sophisticated methods have to be used. Quantum mechanical methods (QM) have to be applied (Chapter 2.3) for the description of charge transport and charge transfer (CT) in molecules.

2.1 Molecular Mechanics

The description of all properties of molecules with methods based on the relativistic time-dependent Schrödinger equation is very difficult, if not impossible, and very time consuming. One has to think about methods which will be much faster without losing too much of the desired accuracy. The properties of biomolecules depend mainly on their atomistic structure, which can be described with force field based methods very well.

To describe the atomistic structure, MM methods condense the electronic structure with the nuclei and therefore depict the atoms in a classical manner. The corresponding energy of such a classical system is due to the bonded and the non-bonded interactions between the atoms. Atoms connected with each other with covalent bonds form certain angles and dihedrals, which are assigned potential energy terms that contribute to the total energy. Atoms that are not bound to each other directly interact in terms of the Coulomb interaction and the van-der-Waals forces. The energy is calculated from all these different bonded and non-bonded interactions. Equation 2.1 shows the total energy of an MM system.

$$
\begin{aligned}
E \;=\;& \frac{1}{2} \sum_{bonds} k_b (r_b - r_b^0)^2 + \frac{1}{2} \sum_{angles} k_a^\theta (\theta_a - \theta_a^0)^2 \\[2mm]
&+ \frac{1}{2} \sum_{dihedrals} V_n \left[1 + \cos(n\omega_n - \omega_0) \right] \\[2mm]
&+ \sum_{i}^{N-1} \sum_{j=i+1}^{N} \left[4\epsilon_{ij} \left[\left(\frac{\sigma_{ij}}{r_{ij}}\right)^{12} - \left(\frac{\sigma_{ij}}{r_{ij}}\right)^{6} \right] + \frac{q_i q_j}{4\pi\epsilon_0 r_{ij}} \right]
\end{aligned}
\tag{2.1}
$$

Some parameters need to be determined in advance to calculate the total energy for the molecule. For all different bond types such as C-C or C=C, which are present in the molecule, a force constant k_b and an equilibrium bond length r_b^0 has to be supplied. The same is true for the force constants (k_a) and equilibrium values (θ_a^0) of all different angles. Three sets of parameters are necessary, the multiplicity (n), the corresponding energy (V_n) and the phase shifts (ω_n^0) for the description of dihedral angles.

The non-bonded interactions are the electrostatic Coulomb interaction of the atom-centered charges and a combination of the van-der-Waals interactions with the Pauli repulsion. For the former, the charges q_i of the atoms are obtained with a fit of the electrostatic potential of the whole molecule obtained from QC calculations.[50] The latter can be described efficiently and with sufficient accuracy by the Lennard-Jones (LJ) potential. Here, the parameters ϵ_{ij} and σ_{ij} describe the depth and location of the LJ potential. All these parameters are fitted to QM calculations or experimental results and tabulated in advance for all relevant atoms and their different binding situations.

Calculating the non-bonded parameters for all pairs of atoms in a huge biomolecule might slow down the calculation significantly. To resolve this issue, simple cut-off schemes, like those used for the van-der-Waals interactions, or more sophisticated methods, like the particle-mesh Ewald (PME) method [51–53] for the electrostatics, can be used to make the computation of the non-bonded terms more efficient.

With initial coordinates and a well-defined force field, the total energy of a molecule and gradients are calculated. Therefore, it is possible to perform molecular dynamics (MD) simulations.

2.2   Molecular Dynamics Simulation

Molecules and atoms are not fixed on their equilibrium positions at finite temperatures. Rather, atoms move and bonds oscillate on certain timescales. Thus, it is not reasonable to investigate only their static, energy-minimizing conformations, especially when dealing with large biomolecules. Molecular dynamic (MD) simulations are performed to get an idea how molecules behave in a defined time frame. The basic idea is to treat the motions of atoms in a classical manner in order to calculate positions $(\vec{r}_i(t))$ and momenta $(\vec{p}_i(t))$ of all atoms in every time step. The result is a trajectory in the phase space of the molecule .

The MD simulation is based on Newton's classical equations of motion [54]

$$\vec{F}_i = m_i \frac{\delta^2 \vec{r}_i}{\delta t^2} \tag{2.2}$$

where $\vec{F}$ is the force acting on the atom i and $\vec{r}$ is the three-dimensional coordinate of the atom i in the system. As a result, a system consisting of N atoms is described by $3N$ coordinates.

The forces are calculated as the negative gradient of the energy.

$$\vec{F}_i = -\frac{\delta E(\vec{r}_i)}{\delta \vec{r}_i} \tag{2.3}$$

An MD simulation can be performed with these equations of motion. These are ordinary differential equations of second order. Numerical methods are available to solve these.

2.2.1   Solving the Equations of Motion

There are several methods for the integration of the equations of motion. Ranging from very simple ones, like the first-order Euler method, to more complicated ones, like the second-order leap-frog method, to even more complicated methods of higher orders. All of these share one fundamental approach, the introduction of a time step. This is necessary due to the fact that the integration can only be done numerically.

The leap-frog method [55] is used in this work. Here, the positions and velocities are evaluated in an alternating fashion.

$$\vec{r}(t),\ \vec{v}(t + \frac{1}{2}\Delta t),\ \vec{r}(t + \Delta t),\ \vec{v}(t + \frac{3}{2}\Delta t),\ \vec{r}(t + 2\Delta t)...$$

The first step is to calculate the velocities $\vec{v}\left(t + \frac{1}{2}\Delta t\right)$

$$\vec{v}\left(t + \frac{1}{2}\Delta t\right) = \vec{v}\left(t - \frac{1}{2}\Delta t\right) + \frac{\vec{F}(t)}{m}\Delta t \tag{2.4}$$

and the new atomic positions are calculated in the next step

$$\vec{r}\left(t + \Delta t\right) = \vec{r}\left(t\right) + \vec{v}\left(t + \frac{1}{2}\Delta t\right)\Delta t \tag{2.5}$$

This algorithm produces trajectories that are identical to those obtained with the Verlet algorithm [56].

Now, a time step has to be chosen for the numerical integration. The value of this crucial parameter in the simulation depends on two factors basically:

- First, there are numerical errors due to the second order Taylor expansion. Contributions in Δt^3 and higher orders are neglected, introducing an error of the same order. To avoid this issue, the step size can be chosen very small, which makes simulations on the desired time scales computationally more demanding

- Second, fast motions in the system have to be accounted for. The fastest motion in a biomolecular system is the C-H bond stretching with a period of about 10 fs. A rule of thumb recommends the step size to be one order of magnitude smaller than this fastest motion. This normally leads to a time step of 1 fs in MD simulations of biomolecules.

Biomolecular processes, like protein folding, happen on the nano- to microsecond or even longer time scales typically. Millions of MD integration steps have to be performed to reach these time scales, which makes the computational efficiency of every time step very important. The calculation of forces takes the larges amount of the computational time (99% ore even more) in every time step. There are more sophisticated integration methods which allow a larger time steps without increasing the numerical error, but on the other hand, more computational time is needed per step.

Therefore, an MD simulation is possible with a chosen integration method and a given force field. In order to perform simulations resembling the experimental setup or even the natural environment, thermodynamic ensembles are introduced.

2.2.2 Thermodynamic Ensembles

The conceptually simplest ensemble is the micro-canonical ensemble (NVE) where the number of particles (N), the volume of the simulation box (V) and the total energy (E) are held constant. As the total energy is conserved, the temperature in the system can be very different in place and time. This is not the situation in a physical system in equilibrium with the environment. To achieve this, temperature and pressure couplings are introduced.

Temperature Coupling

The canonical ensemble coupled to an external heat bath is called NVT ensemble. A common way to describe the bath is the Berendsen thermostat [57]. Any deviation from the reference temperature is corrected between two time steps depending on the pre-set time constant τ.

$$\frac{dT}{dt} = \frac{1}{\tau}(T_{target} - T)$$ (2.6)

Thus, we can calculate the change in temperature

$$\Delta T = \frac{\Delta t}{\tau}(T_{target} - T)$$ (2.7)

as there is a direct relation between the temperature and the velocities of the particles in the system.

$$\frac{1}{2}m \left\langle v^2 \right\rangle = \frac{3}{2}kT$$ (2.8)

The temperature in a system is changed by scaling the velocities of all particles in the system with a scaling factor λ, which is obtained from the relation

$$\Delta T = \frac{\Delta t}{\tau}(T_{target} - T) = (\lambda^2 - 1)T$$ (2.9)

as

$$\lambda = \sqrt{1 + \frac{\Delta t}{\tau}\left(\frac{T_{target}}{T} - 1\right)}$$ (2.10)

The coupling to the external heat bath now depends on the choice of the parameter τ. With higher values for τ, the temperature will converge faster to the reference temperature T_{target}.

Unfortunately, the Berendsen thermostat does not yield a correct canonical probability distribution. A more sophisticated thermostat that rigorously represents a correct canonical ensemble is the widely used Nosé-Hoover thermostat [58, 59]. Here, a heat reservoir is introduced as an integral part of the system with one degree of freedom s, to which a mass Q is associated.

Now, the equations of motion have to be calculated for an extended system with 3N+1 degrees of freedom

$$\ddot{r}_i = \frac{F_i}{m_i} - s \cdot \dot{r}_i \tag{2.11}$$

where the second term can be seen as a kind of a "friction" originating from the heat bath.

Then, the extra equation of motion for the heat bath takes the following form:

$$\dot{s} = \frac{1}{Q}(T - T_{target}) \tag{2.12}$$

The strength of coupling is determined by the introduced parameter Q, the "mass" of the heat bath. To get a more intuitive parameter for the coupling, a time parameter τ is introduced.

$$Q = \frac{\tau^2 \cdot T_{target}}{4\pi^2} \tag{2.13}$$

The period of oscillation of the kinetic energy between the system and the heat bath is specified by this parameter τ.

The thermostat is incorporated into the equations of motion directly in the Nosé–Hoover approach. Therefore, it is an integral part of the integrator and not only an *a posteriori* correction. It can be shown that the ensemble generated by the Nosé–Hoover thermostat is a correct canonical ensemble.

Pressure Coupling

Chemical experiments are normally conducted under conditions of constant pressure. An NPT ensemble is used to describe such conditions in MD simulations. In contrast to the thermostats where the velocities are subject to scaling, the barostats scale the volume of the simulation box. This can be done equivalently to the methods described in the previous part. The Berendsen barostat scales the volume of the box in a direct fashion and thus shares the same drawbacks with the Berendsen thermostat. Again, a more sophisticated way is to introduce an additional degree of freedom and solve the equations of motion directly with the barostat incorporated. In this sense, the Parrinello–Rahman barostat [60, 61] is the equivalent to the Nosé–Hoover thermostat.

2.3 Quantum Chemistry

MM methods, like shown in the latter chapter, are only able to describe the structure and dynamic properties of molecules without accounting for their explicit electronic structure. In contrast, quantum chemistry (QC) methods are designed to describe the electronic structures of molecular systems. Here, a short introduction into the basic concepts of the QC methods, which are applied in the studies of CT in the later parts of this work, will be given. The starting point will be the density functional theory (DFT). Having summarized the DFT, the approximative density-functional tight-binding (DFTB) (chapter 2.3.2) method is introduced. The latter will be used in the CT calculations.

2.3.1 Density Functional Theory

DFT is a quantum chemical method to describe electronic structures of atoms and molecules. In contrast to the wave function based methods like Hartree–Fock (HF) and post-HF (e.g. perturbation theory or configuration interaction), DFT is based on the calculation of the electron density. This approach has two main advantages:

- The system is no longer described by 3N coordinates of the wave function, but only by 3 coordinates overall.

- The electron density is a physical observable whereas the wave-function (WF) is not.

The probability of finding an electron at a position r can be computed by integrating over all other electrons in the system.

$$\rho(r_1) = N \int \Psi^*(r_1...r_N)\Psi(r_1...r_N)dr_2...dr_N \tag{2.14}$$

However, this is not the way to go, since the determination of the true N-particle wave function is a complicated task. Therefore, a less complicated approach is sought which determines the density directly and is able to calculate the energy of the system directly from this density.

An important point here is to prove that the density is an unique feature of a certain system and therefore the energy can be uniquely determined. Moreover, the obtained density functional $E[\rho(r)]$ has to follow the variational principle for energy minimization. In the following, both of these issues were solved.

It is well established that the electron density is uniquely determined in a given external potential $V_{ext}[\rho(r)]$. To use a density-functional for the description of system properties this correlation has to be invertible. Hohenberg and Kohn proved in 1964 that the external potential and thus the ground state properties of a system are determined uniquely by the electron density $\rho(r)$ [62].

$$\rho(r) \rightarrow V_{ext}[\rho(r)] \tag{2.15}$$

The second theorem by Hohenberg and Kohn implies that the energy functional takes a minimum value for the ground state density:

$$E[\rho_0] \leq E[\tilde{\rho}] \tag{2.16}$$

where ρ_0 is the electron density at the ground state and $\tilde{\rho}$ is an arbitrary density.

Additionally, the Born–Oppenheimer approximation [63] is applied. Due to the difference in mass between the electrons and nuclei, the motions of these particles can be treated separately.

With these issues taken care of, a density-functional can be derived and the total energy of the system takes the form:

$$E[\rho(r)] = T[\rho(r)] + V_{ext}[\rho(r)] + V_{ee}[\rho(r)] + E_{xc}[\rho(r)] + J_{NN}[\rho(r)] \tag{2.17}$$

where $T[\rho(r)]$ is the kinetic energy of the electrons, $V_{ext}[\rho(r)]$ is the energy of the electrons in the external field of the nuclei, $V_{ee}[\rho(r)]$ is the energy from the classical electron-electron interaction, $E_{xc}[\rho(r)]$ is the exchange correlation energy missing in the classical term and $J_{NN}[\rho(r)]$ is the interaction between the nuclei.

While the functionals for the electron-electron and the electron-core interaction are known from the Hartree theory, the exact form of the functionals for the kinetic energy of the electrons and the exchange correlation are unknown.

In 1965, Kohn and Sham introduced the idea to describe the kinetic energy by using Hartree–Fock like orbitals, the so-called Kohn–Sham orbitals [64].

In this approach, an imaginary set of one-electron equations, the Kohn–Sham equations, is built.

$$\left(-\frac{1}{2}\nabla^2 + V(r) \right) \phi_i = \epsilon_i \phi_i(r) \tag{2.18}$$

Although, this construction counteracts the idea of orbital-free DFT, it is until now the only computationally convenient and widely used solution.

The still missing $E_{xc}[\rho]$ now has to be approximated. Many methods have been developed, and the most widely used are:

- local density approximation (LDA)

$$E_{xc} = F[\rho(r)]$$

- generalized gradient approximation (GGA)

$$E_{xc} = F[\rho(r), \nabla\rho(r)]$$

- meta-GGA

$$E_{xc} = F[\rho(r), \nabla\rho(r), \nabla^2\rho(r)]$$

- hybrid functionals: e. g. adding HF exchange (e.g. B3LYP)

With these methods, the electronic structure of any system can be determined. Despite the idea of only using the density, the required orbitals for the calculation of the kinetic energy can make DFT based methods very slow. Therefore, in the following section an approximative DFT method is introduced, which is able to describe systems of relevant sizes on a quantum mechanical level more efficiently.

2.3.2 Approximative DFT – Density-Functional Tight-Binding

Nowadays, DFT is widely used for the calculation of properties of molecular systems of up to hundreds of atoms. One disadvantage of DFT is the large amount of computational time needed for the evaluation of integrals during runtime. Especially when calculating electronic structures along MD trajectories, DFT becomes too expensive in computational cost. To avoid this issue, there are many semi-empirical approaches, which achieve higher computational efficiency by neglecting contributions deemed insignificant or parameterizing integrals in advance. One way to achieve this is presented in the following [65].

Density-functional tight-binding (DFTB) is based on a minimal basis set, thus describing only the valence electrons of atoms, while the inner electrons are treated by two-center potentials. Thereby, crystal-field and three-center terms are neglected. The two-center Hamiltonian and overlap matrix elements are tabulated for distances of up to a reasonable threshold. This way, none of the integrals have to be computed at runtime. Rather, they are interpolated for the current distance between the atoms. Finally, the DFT double-counting terms are included into the repulsive energy term. This term is pre-calculated and tabulated as well. In the parametrization, the values for the repulsive energy are calculated from the difference of the energy calculated with DFT and the tight-binding electronic atomization energy for chosen representative molecules. With these parametrizations DFTB is up to 3 orders of magnitude faster than DFT.

The starting point for DFTB is to construct a reference density ρ_0 from a superposition of densities calculated on isolated atoms.

$$\rho_0 = \sum_{\alpha} \rho_\alpha \tag{2.19}$$

This density will differ from the DFT density by $\Delta\rho$. The total energy then can be calculated by:

$$E[\rho] = E[\rho_0] + O(\Delta\rho^2) \tag{2.20}$$

Assuming the reference density ρ to be sufficiently close to the ground state density ρ_0, the Kohn–Sham equations can be solved non-self-consistently and the Kohn–Sham eigenvalues ϵ_i are obtained.

$$E_{elec}^{occ} = \sum_i \epsilon_i \tag{2.21}$$

This represents the contribution of the Kohn–Sham eigenvalues to the total energy of the system. This energy is called electronic energy. To obtain the total energy, another term, called the repulsive energy, which incorporates all other missing contributions is needed.

$$E[\rho] = \sum_i \epsilon_i + E_{rep} \tag{2.22}$$

Two-body terms $V_{\alpha\beta}$ are introduced to represent these repulsive contributions, being exponential functions fitted to reproduce geometries, vibrational frequencies and reaction energies.

$$E[\rho] = \sum_i \epsilon_i + \frac{1}{2}\sum_{\alpha\beta} V_{\alpha\beta} \tag{2.23}$$

Already at this point, DFTB is able to describe unpolar systems, where no charge transfer happens between the single atoms. As many hetero-nuclear molecules, including biomolecules, have differences in electronegativity, a more sophisticated approach has to be sought. The self-consistent-charge DFTB method (DFTB2) overcomes this issue by a second-order Taylor expansion of the DFT total energy. Here, charge density fluctuations $\Delta\rho$ around a given reference density ρ_0, which is the superposition of neutral atomic densities, are considered[66–68].

The density differences $\Delta\rho$ are now approximated as atomic contributions:

$$\Delta\rho = \sum_\alpha \Delta\rho_\alpha \qquad (2.24)$$

Now, the charge monopole approximation is used for the atomic density functions.

$$\Delta\rho_\alpha \approx \Delta q_\alpha F_{00}^\alpha Y_{00} \qquad (2.25)$$

Here, F_{00}^α is the normalized radial dependence of the density fluctuation on atom α which is approximated to be spherical (Y_{00}).

The resulting approximated second order energy terms has two limiting cases:

1. When atoms α and β are at a large distance (>2 Å), the exchange correlation energy becomes zero. In this case, the integral only describes the Coulomb interaction of two point charges.

$$E^{2nd} \approx \frac{1}{2} \sum_{\alpha\beta} \frac{\Delta q_\alpha \Delta q_\beta}{R_{\alpha\beta}} \qquad (2.26)$$

2. For $\alpha = \beta$, the integral describes the electron-electron interaction on a single atom α

$$E^{2nd} = \frac{1}{2} U_\alpha \Delta q_\alpha^2 \qquad (2.27)$$

The parameter U_α is the Hubbard parameter newly introduced in DFTB2 and describes the chemical hardness of the atom α. It is a measure for the energy change of a system upon addition or removal of electrons.

In conclusion, this introduced method is an approximative DFT method which is up to 3 orders of magnitude faster than DFT. This is achieved by using a minimal basis set for the valence electrons of every atom as well as pre-calculating and tabulating two-center integrals for all atom pairs. Therefore, DFTB2 is able to describe large systems like DNA over extended periods of time and is the method of choice for the charge transfer calculation in this work.

2.4   Dynamics of Excess Charge in DNA

In spite of excessive research in the field of charge migration through biomolecules, especially through DNA, no unified explanation of the microscopic mechanisms has been found until now. Two basic concepts regarding the dynamics of charge have been established, the charge transport and the charge transfer. The former, is the physicist's approach which describes the ability of DNA to support electric current. The latter is the chemical (physics) approach of charge transfer in DNA involving multi-step hopping processes.

Either way, the computational model used for charge transport/transfer description in DNA has to fulfill several requirements. Firstly, the computational efficiency has to be high enough to take dynamic effects on the nanosecond timescale into account. Secondly, it has to be able to reproduce the results of higher-level methods in a qualitative way.

The first concept used in this work will be the calculation of charge transport properties with QM methods on structural ensembles originating from MD simulations. In this case, the molecular structure is simulated without the charge actually being in the system. Thus, the structure and dynamics of the molecule are unaffected by the charge in the system. The CT properties are calculated on these "neutral" structures over several thousand snapshots. This provides an insight into the dynamics of CT efficiency. This previously developed multi-scale framework for the dynamic calculation of charge transport properties [45] has already been applied successfully to fully hydrated DNA species in a free state.[44, 69, 70]

In the second part, mechanisms of charge transfer will be introduced. The classical description of charge transfer in molecular systems in a polarizable medium goes back to the theory of Marcus [71]. First, the general concept and the needed CT parameters will be introduced. Then, the computational approaches used in this work will be presented. The computation of charge transfer here is based on a direct combination of force field and quantum methods. The MM environment is influenced by the charge residing on the nucleobases and reacts accordingly. To make such an approach computationally efficient, several assumptions have to be made, which will be described in detail. The basics of this multi-scale approach have recently been described by T. Kubař and M. Elstner[72, 73] and will be summarized in this thesis.

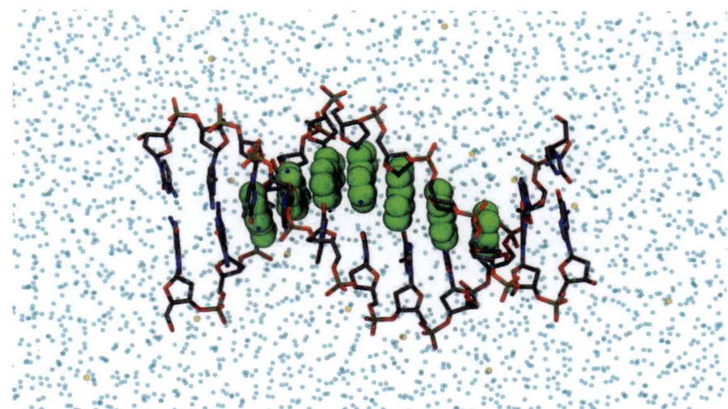

Figure 2.1: *Typical QM/MM system of DNA in water environment. The water molecules (blue spheres), the ions (orange spheres) and the backbone of the DNA are calculated classically with force field methods. The purine nucleobases involved in the CT process (shown in green) are treated quantum chemically.*

2.4.1   The Multi-Scale Framework

The general issue of computation of CT in large systems is that QC methods are needed to describe the dynamics of the wave function of the excess charge. High-level electronic structure calculations can only describe tens of atoms over short periods on the picosecond timescale. QC methods like DFT are more computationally efficient and therefore can describe larger systems up to hundreds of atoms, but still only for very short periods of time. So, a reasonable approach in this quest for computational efficiency is to use approximative QC methods like DFTB2, which are able to describe hundreds of atoms on the nanosecond time scale. Still, for large molecular systems like DNA in a solvent or a whole protein, the part of the system which is described with these QC methods has to be restricted. A sophisticated solution to this is a QM/MM scheme where a small QM region is defined and treated with QC, while the rest of the system is treated classically with MM force field methods.

Figure 2.1 shows a DNA double helix in water environment. The water, the ions and the DNA backbone are treated classically, while the purine bases (shown in green) are treated with QM methods.

This way, the QM region is being influenced by the MM environment and vice versa. See figure 2.2 for an overview of the computational framework. Shown are the QM and MM calculation steps and the parameters communicated between

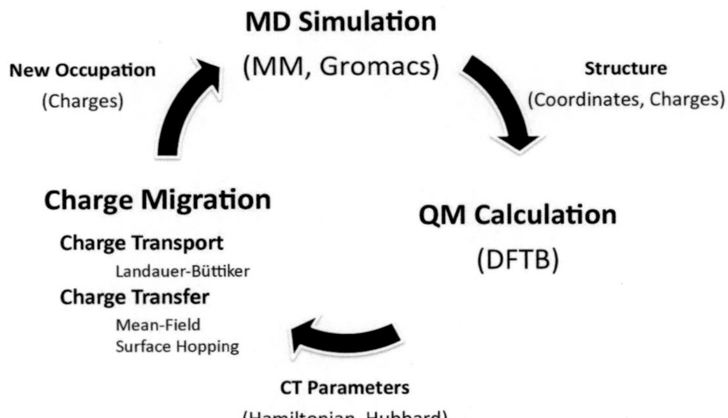

Figure 2.2: *Overview of the steps in the multi-scale approach. The MM simulation generates the structure of the molecule. This is passed to the QM calculation, performed with DFTB2, which generates the Hamiltonian. This Hamiltonian can be used for charge transport calculations with the Landauer–Büttiker theory. Or, the wave-function is propagated by the the mean-field (Ehrenfest) or the surface hopping method. Finally, the atomic charges in the MM system are updated according to the wave-function.*

the calculations. This interaction is crucial for the description of charge transfer in biomolecular systems and cannot be neglected. The next step in this approach is to reduce the necessary amount of computation time for the still large QM region. Charge transfer is supposed to occur only via the frontier orbitals of the involved molecules. In the case of hole migration in DNA, only the highest-occupied molecular-orbitals (HOMO) of the purine bases are treated explicitly. This can be seen as a fragment-orbital approach where a part of the electronic structure is frozen and treated implicitly like the σ and π separation in the Hückel theory [74, 75].

2.4.2   The Fragment Orbital Approach

To achieve a reasonable computational efficiency, a QM/MM model is introduced where only the excess charge is calculated with QM methods while the rest of the molecule is treated with classical MM methods. The general idea is to decompose the system into M spatially non-overlapping fragments. With this assumption, it is possible to perform separate QM calculations of every fragment. This procedure may be called a "fragment molecular orbital approach" (FMO), which was originally developed by Kitaura et al. [76]

The fragment orbitals ϕ are linear combinations of atomic orbitals.

$$\phi_m^i = \sum_\mu c_\mu^{im} \chi_\mu \tag{2.28}$$

where c_μ^{im} are the FMO expansion coefficients and χ are the atomic orbitals. These FMO can be obtained with QM methods. Each of the FMO is localized on one single fragment by definition and these calculations can be done separately for every fragment. But, it is of great importance to include the electrostatic environment into these calculations, as the FMO are strongly influenced by them. The implementation used in this work uses the DFTB2 method to perform a SCF calculation for every fragment, a nucleobase in the case of DNA, in every time step. The electrostatic environment of the fragment is included as point charges.

Now, the FMO are considered to constitute the basis for the expansion of the wave-function of the excess charge. In case of hole transfer, the wave-function is built from the HOMO of the single fragments.

$$\Psi(r, R) = \sum_m a_m \phi_m^{HOMO} \tag{2.29}$$

The obtained FMO are used to build a Hamilton matrix.

$$H_{mn} = \langle \phi_m | \hat{H} | \phi_n \rangle = \sum_\mu \sum_\nu c_\mu^m c_\nu^n \bar{H}_{\mu\nu} \tag{2.30}$$

where indices μ and ν run over the relevant atomic-orbital-like basis functions of the fragments m and n, and $\bar{H}_{\mu\nu}$ is the Hamiltonian in this atomic-orbital basis. This Hamilton matrix can be set up easily. The fragments m and n are calculated separately to determine the c_μ^m and c_ν^n, while the $\bar{H}_{\mu\nu}$ is directly read in from the pre-calculated and tabulated DFTB2 parameters.

Finally, the set of FMO is orthogonalized to simplify the following propagation of the wave-function.

In the simplest approach to propagate this wave-function now is to perform a Born–Oppenheimer simulation. Here the Hamiltonian is diagonalized to yield the wave-function coefficients a_m directly.

But, the wave-function can also be propagated with more sophisticated non-adiabatic schemes, which will be described in the following chapter 2.6.

Application to DNA

The part of the DNA which is the easiest to oxidize are the nucleobases, particularly the purines adenine (A) and guanine (G). Guanine has the lowest ionization potential of all nucleobases while the IP of adenine is 0.4 eV higher[77, 78]. Since cytosine (C) and thymine (T) have an ionization potential about 1 eV higher, G and A are supposed to be the charge carrying sites in DNA in case of hole transport/transfer. Therefore, the hole transfer in DNA in this model occurs between the HOMO of guanine and adenine nucleobases, which form a linear chain of charge carrying sites.

Issues with this QM/MM approach

This multi-scale approach suffers from certain issues. Some of them are resulting from the assumptions made, and some are inherent to the applied methods.

Firstly, the treatment of the single fragments involves approximations that require discussion. When the point charges of the MM system are updated after a step of dynamics, the Coulomb interactions of those atoms do change. But, these interactions are not evaluated for neighbors and next-neighbors in the MM calculations. This way, the change of Coulomb interactions is neglected nearly completely. Moreover, the parameters of the bonds and angles do not change either as they are fixed values of the used force-field. This way, the inner-sphere reorganization energy (RE, see also chapter 2.6.1) is neglected completely. To account for this essential flaw, the inner-sphere RE is introduced as an empirical contribution to the energy of the system. Quantum chemical calculations yield a value for λ_i of about 0.23 eV for adenine and guanine [79].

Secondly, the interactions between the single fragments should change when the point charges are updated. Here, the Coulomb interactions are covered by the MM calculation. What is missing is the change in the van-der-Waals parameters of the fragments. These are fixed values in a force-field method and do not change during a simulation. As these effects are very small, it is likely that this effect is negligible.

And finally, the polarization of the environment is not covered completely by the reorientation of the molecules alone. The electric-charge density is represented by the MM point charges, while the electronic polarization is not considered in the force-field used. This is a crucial point, because the fluctuations of the IP of the fragments are overestimated in such a non-polarizable force-field. As an efficient and reliable implementation of a polarizable force-field is not available, a simple scaling of the electronic interaction is performed to reduce the energy barriers of CT. In the used TIP3P water environment, these interactions are overestimated by 26-34% [80], therefore QM/MM interactions are scaled down by a factor of 1.5 [73].

2.5 Charge Transport in DNA

2.5.1 Landauer–Büttiker Framework

The method for charge transport calculations used in this work is based on the Landauer–Büttiker framework. This method represents the physicist's view of a coherent charge transport through a system. The CT parameters are obtained with the presented multi-scale framework and used to calculate transmission functions for single MD snapshot structures. To account for dynamic fluctuations, thousands of snapshots from MD simulations will be averaged over. With this setup, an insight into the dynamic charge transport properties of the investigated systems will be obtained.

As described previously, only the highest occupied molecular orbitals (HOMO) of the purine bases were considered as states participating in hole transport. While this choice is a justified first approximation, the application of higher voltage could make other states accessible, for instance lower occupied orbitals of the purines or orbitals of the pyrimidines [81, 82] Note further that transport of excess electron is not taken into account with this approach.

The transmission function can be calculated using the Greens function with the Fisher–Lee relation,[83]

$$T(E) = |G_{1N}(E)|^2 \tag{2.31}$$

$G_{1N}(E)$ is the $(1, N)$ element of the chain Green's function matrix, which was calculated via the matrix Dyson equation,

$$\mathbf{G}(E) \;=\; (E\mathbf{1} - \mathbf{H})^{-1} \tag{2.32}$$

Within the Landauer–Büttiker framework it is possible to calculate an instantaneous current. To do so, some additional assumptions have to be made. First, the chain of charge carrying sites has to be coupled to an electrode on each side.

The transmission then contains two additional terms, which account for this coupling.

$$T(E) = 4\gamma_L \gamma_R |G_{1N}(E)|^2 \tag{2.33}$$

The wide-band-approximation is assumed here, neglecting any atomic structure of the electrodes. Then, the coupling to the electrodes is described by single values γ_L and γ_R as effective coupling terms. Therefore, the molecule is considered to be contacted to a perfect metal with a constant, energy-independent density of states. Note here that the assumption of constant coupling to the electrodes may be unrealistic when stress is exerted upon the DNA–electrode contact, which makes the contact loosen, transform or even break. In this case the $(1, N)$ element of the chain Green's function matrix $G_{1N}(E)$ contains the coupling to the electrodes,

$$
\begin{aligned}
\mathbf{G}(E) &= (E\mathbf{1} - \mathbf{H} - \mathbf{\Sigma}_{\mathrm{L}} - \mathbf{\Sigma}_{\mathrm{R}})^{-1} \\
(\mathbf{\Sigma}_{\mathrm{L}})_{lj} &= -i\,\gamma_{\mathrm{L}}\delta_{l1}\delta_{j1} \\
(\mathbf{\Sigma}_{\mathrm{R}})_{lj} &= -i\,\gamma_{\mathrm{R}}\delta_{lN}\delta_{jN}
\end{aligned}
\tag{2.34}
$$

Then, the instantaneous current can be obtained with the integration of the transmission function as

$$
I(V) = \frac{2e}{h}\cdot 4\gamma_{\mathrm{L}}\gamma_{\mathrm{R}}\cdot \int_{-\infty}^{\infty} \mathrm{d}E\ \left(f\left(E - \frac{eV}{2}\right) - f\left(E + \frac{eV}{2}\right)\right)\ |G_{1N}(E)|^2
\tag{2.35}
$$

with f being the Fermi–Dirac distribution function.

To obtain single values for the instantaneous current, the converged limiting current at the voltage of 2 V, which is independent of the choice of the Fermi level, was considered.

With the application of this framework, two further conditions are assumed:

- the CT is purely coherent, meaning that possible incoherent scattering would be missed [84]

- the motion of electrons is faster than the motion of nuclei.

These issues were addressed in Refs. [44, 45], where the methodology was described in detail. Note that the aim here is not to compute quantitative current–voltage characteristics, for which explicit evaluation of Fermi energies and account of all states possibly contributing to conductance at the considered voltage would be required. Rather, the maximum current at high voltage is considered as a convenient measure of the ability of the DNA-based molecular system to support coherent CT. Actually, the CT parameters (electronic coupling (EC) and ionization potential(IP)) are the key quantities determining the rate of CT proceeding with any mechanism. Therefore, the obtained values of maximum current can be considered to be indicative for the efficiency of CT in general.

2.6 Charge Transfer in DNA

In this chapter, a short introduction into the basics of charge transfer and the needed parameters is given. Then, two advanced methods for the propagation of the wave-function of the excess charge will be summarized. This represents the rather chemical point of view of a charge transfer process, where the charge resides on one molecule, or part of the molecule, and then is transferred to another molecule. Two of these non-adiabatic methods were implemented in our framework and recently published[72, 73].

2.6.1 Basics of Charge Transfer

The classical way to the CT in complex molecular systems is Marcus' theory [71, 85] and its extensions [86–90] Here, a rate of transfer between a donor and an acceptor for the non-adiabatic case can be calculated from several charge transfer parameters.

$$k_{DA} = \frac{|V_{DA}|^2}{\hbar} \sqrt{\frac{\pi}{\lambda k_B T}} exp \left[-\frac{(\Delta G^0 + \lambda)^2}{4\lambda k_B T} \right] \tag{2.36}$$

These parameters are:

- the electronic coupling (EC) $|V_{DA}|^2$ between the donor and the acceptor

- the reaction free energy ΔG of the CT process

- and the reorganization energy (RE) λ

All three can be obtained using various computational methods. The static electronic couplings can be calculated with highly accurate quantum chemical (QC) methods like CAS-PT2. More efficient methods have to be used to calculate the dynamic electronic couplings in the course of an MD simulation. In our approach, the dynamic EC are obtained using the DFTB2 method (see section 2.3.2 for details) and a fragment orbital approach[91]. ΔG and λ are thermodynamic quantities, so that their computation can be done by sampling of the configuration space, e.g. free energy calculations. The RE is the energy needed to reorient the molecules and their environment to the conformation corresponding to the final state of the CT reaction.

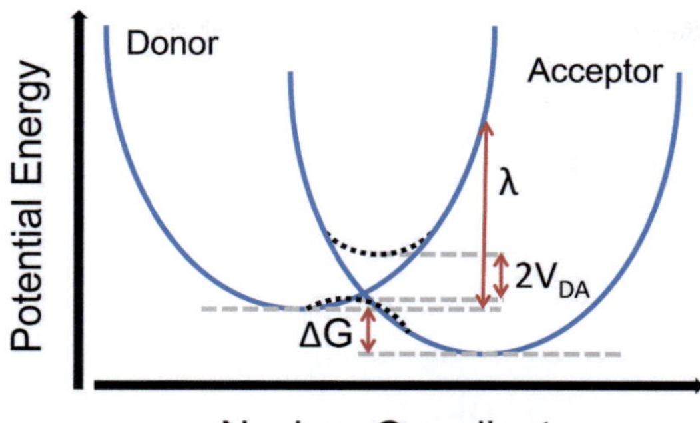

Figure 2.3: *Energetic profiles of donor and acceptor system in the Marcus picture. The solid lines show the (diabatic) ground state of the system. Dashed lines show the adiabatic states which occur with the avoided crossing of the two potentials. Denoted are the three fundamental CT parameters.*

The first part is the inner-sphere RE (λ_i) which accounts for the structural changes in the charged molecule. The second contribution is the the outer-sphere RE (λ_s) which is the energy needed to reorient the environment of the charged molecule. The inner-sphere RE has to be computed with quantum-chemical methods[92], while the outer-sphere RE of the environment (e.g. the solvent) can be calculated using MM methods. It is extremely important to achieve high accuracy in these parameters as they occur in the exponential term of the Marcus' equation. Especially, the computation of the outer-sphere RE is prone to errors as it is highly force-field dependent and overestimated in general by non-polarizable force-fields [80].

Figure 2.3 shows the general situation in the Marcus' picture. The energy of both charged states is approximated by a parabola. The meaning of the three parameters can be illustrated by this qualitative picture.

The general issue with Marcus' theory is the several introduced assumptions. First, the electronic structure of the donor and acceptor system has to be known well for the charged and for the neutral state. This includes knowledge about the (de-)localization of the charge in the system. Second, the CT has to be significantly slower than the structural dynamics of the system. And finally, the mechanism of transport has to be known (adiabatic, non-adiabatic, solvent-controlled, etc.)

To overcome these limitations, more advanced charge transfer calculation methods were developed and are used in our laboratory. These methods are free of the assumptions mentioned in connection with Marcus' theory.

Computing the Reorganization Energy

The RE of CT processes of molecules in an aqueous solvent is mainly influenced by the outer-sphere RE introduced through the solvent. In chapter 6 this parameter will be investigated in detail. Therefore, the method used to compute this parameter is introduced here.

The previously developed MD-based scheme will be used to calculate the outer-sphere RE (λ_s) for CT in DNA [93].

The general idea is to calculate the RE of a charge transfer process from a state α, where the charge is localized on one molecule, to a state β, where the charge is localized on the other molecule, where the environment has to be reorientated.

$$\lambda_s = \langle E_\alpha(r)\rangle_\beta - \langle E_\alpha(r)\rangle_\alpha \tag{2.37}$$

With the index of E denoting the Hamiltonian used to calculate the energy of the system, and the index of the parenthesis denoting the Hamiltonian used to create the ensemble. This allows the estimation of the outer-sphere RE using MD simulations.

$\langle E_\alpha(r)\rangle_\alpha$ is the mean potential energy of the simulation being in state α, calculated with the appropriate Hamiltonian for this state. On the other hand, $\langle E_\alpha(r)\rangle_\beta$ is the mean potential energy of the ensemble, created with the Hamiltonian β, calculated with the Hamiltonian for state α.

2.6.2 Non-adiabatic Propagation Schemes

Mean-Field (Ehrenfest) Theory

The first method used in this work is based on the mean-field (Ehrenfest) [94, 95] approach. Here, the energy of the system is an average over all adiabatic states weighted by their populations. All these states interact with a common potential induced by the classical environment. On the other hand, the classical environment interacts with the combination of these adiabatic states of the quantum system. This way, a single potential-energy landscape is constructed. The propagation of

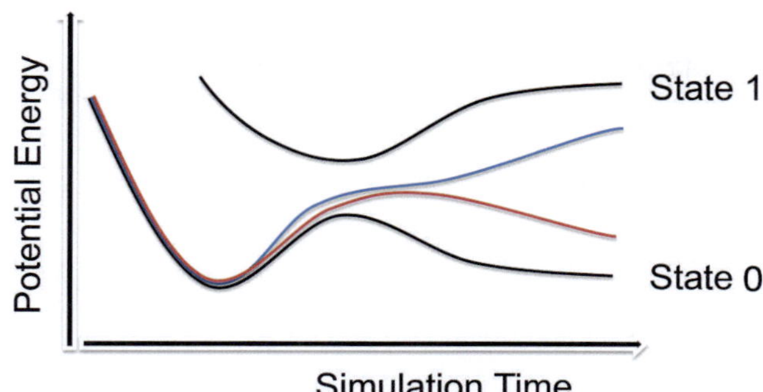

Figure 2.4: *Schematic representation of the behavior of the wave-function in mean-field (Ehrenfest) simulations. The possible course of two simulations is shown. Reaching the high-coupling region, the two states begin to mix and the energy of the actual state remains in between the two adiabatic states. (Based on Fig.5 in [72])*

the wave-function is computed by solving the time-dependent Schrödinger equation (TDSE). The obtained solution is a combination of populated adiabatic states. Figure 2.4 shows the resulting mixed states of two possible mean-field simulations.

The basic implementation works as follows. Two sets of equations of motion are solved simultaneously. First, the ones of electronic system which evolves according to the TDSE. Second, the classical Newton's equations of motion for the MM system. These two sets of equations are coupled to each other. The MM equations contain the charges of the QM system, which are mapped to the MM atomic charges. The TDSE depend on the atomistic structure of the MM system via the Hamiltonian. Therefore the equations have to be propagated simultaneously and this is achieved in a leap-frog like manner. First, the coarse grained QM step is performed and the excess charge is propagated. Then the new charges are mapped to the atomic charges of the MM system. And, a step of classical MM simulation with these atomic charges is performed. After this, the cycle is complete and the next QM step is performed. With this implementation, the classical system feels the actual distribution of charge in the quantum region and can respond to it accordingly. The environment of the charge polarizes and forms a stabilizing polaron. This effect has been found of great importance [96–99].

The presented mean-field approach has some advantages over simpler propagation schemes, like those used before, which had no coupling to the environment and

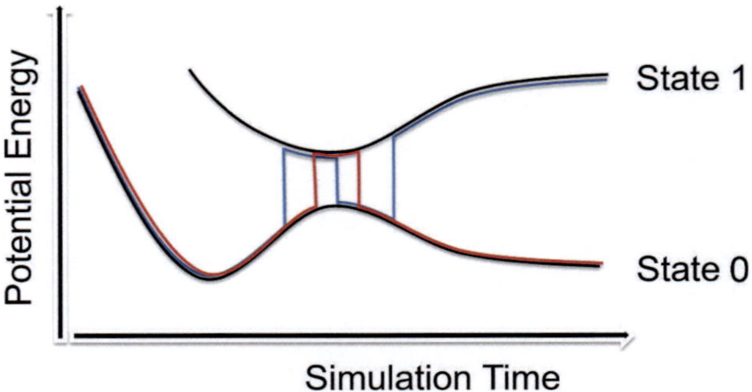

Figure 2.5: *Schematic representation of the behavior of the wave-function in surface hopping simulations. The possible course of two simulations is shown. Reaching the high-coupling region, the two states do not mix like in the mean-field approach. The System may hop between the states and leave the region on one of them (Based on Fig.5 in [72])*

therefore no polarization of the MM environment[100–102]. It is not necessary to choose a representation in advance, without knowing anything about the CT process. The CG Hamiltonian needs not to be diagonalized (as in the following surface hopping method) with this method, which saves computational time. The propagated wave-function can be transferred directly to the classical MD simulation by the mapping to the atomic charges. On the downside, the mean-field approach suffers from the so-called mean-field error. The main issue here is that all adiabatic states interact with the same mean potential of the environment. This has the effect, that in homogeneous DNA sequences, where the energetic landscape is very shallow, the delocalization of the charge is overestimated excessively.

Surface Hopping

The other method of charge transfer computation in this work is the surface hopping (SH) scheme [103]. In this approach, the TDSE is propagated as well, but the classical system interacts only with one pure adiabatic state. The key here is to choose this adiabatic state in every time step. Figure 2.5 shows two possible surface hopping simulations, where hopping occurs in the high-coupling region. The final states are adiabatic ones, no mixing of adiabatic states occurs here.

The propagation of the wave-function then takes place on this adiabatic state through time. Transitions between the states may occur in every time step an can be called hops or switches. The key issue to be solved here is to calculate the probability of such a hop in every time step under the given circumstances.

The implementation used in this work is based on the SH algorithm by Persico et al. [104]. This algorithm uses the local diabatization of the adiabatic states of the system. The advantage of this is a stable representation of adiabatic states in regions of avoided crossing or conical intersections.

In the implemented setup, the eigenproblem of the excess charge is solved in every time step. From this, the wave-function of the excess charge is given as a combination of the adiabatic states. Then the diabatic states are redefined to be identical with the ones at the beginning of the time step. The TDSE is then solved for this locally diabatic basis to propagate the wave-function of the excess charge. The crucial step here is the calculation of the probability to hop from the current state to one of the calculated new states from the populations. The transition to another state is then determined by drawing a random number.

This is an alternative to propagating the wave-function with the mean-field method. The rest of the simulation protocol is identical with that in the mean-field method.

Here again, the propagated wave-function can be transferred directly to the classical MD simulation by the mapping to the atomic charges. The advantage of this setup is that there is no over-delocalization issue like in the mean-field approach. However, in this method the CG Hamiltonian has to be diagonalized. All representations of the quantum system consist of individual states. Therefore it is necessary to choose a representation in advance.

There is no artificial CT when the electronic couplings are very low or even zero. And finally, the microscopical reversibility is fulfilled, again in contrast to the mean-field approach,.

Unfortunately, also the current implementation of the SH method suffers from some issues, which will be outlined shortly. After a surface hop, the velocities of the classical MM atoms are not rescaled. This possibly means that there is no energy conservation. Another issue are the classically forbidden transitions, which are not treated in our setup. The quantum system is supposed to have enough energy at any time for the hop to occur.

CHAPTER 3

Simulation Setup

In this chapter, the developed simulation setup will be described shortly. This work deals with the simulation of the structure and the charge transport characteristics of DNA. Therefore, an introduction into the molecular structure of DNA will be given. Then, the studied DNA species and their designation in this work will be introduced. The simulation setup of the MM calculations is then described in detail to give insight into the specific challenges of the simulation of DNA under experimental conditions. The first subject here is the simulation of DNA under mechanical stress, like it is caused by pulling apart the attached electrodes in the CT experiments. Further on, simulations of DNA strands with reduced water amount are performed. In the end, in this chapter the complete framework of force-field based molecular dynamics simulations for DNA under experimental conditions is presented.

3.1 The DNA Molecule

A DNA strand is composed of monomeric nucleotides which build up to a linear chain like polymer. One of these monomers is composed of a phosphate group and a five-carbon sugar, which build the so called backbone, and a nucleobase.

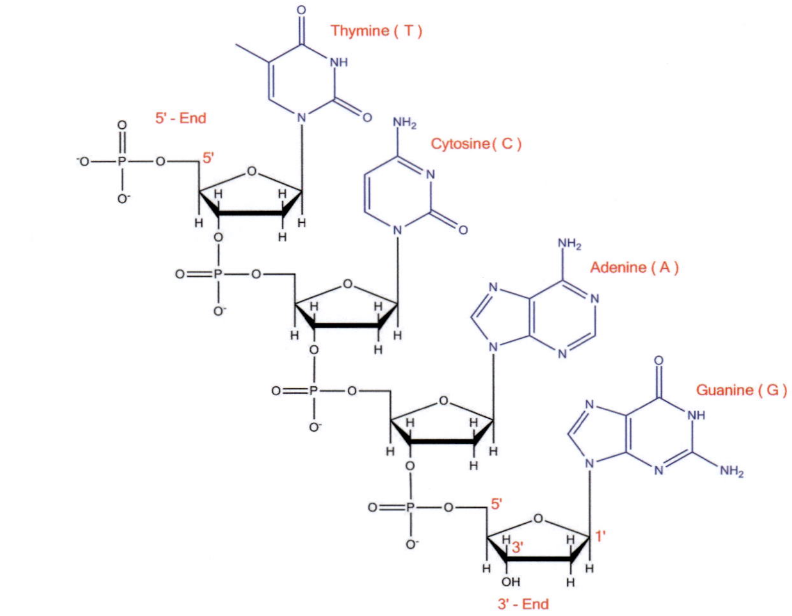

Figure 3.1: *A typical DNA strand composed of the four different nucleotides. Nucleobases are shown in blue, phosphate groups and five-carbon sugar form the backbone and are shown in black. Numbering of the sugar atoms is shown for the lower nucleotide.*

In DNA there are 4 different types of these nucleobases: the purines adenine (A) and guanine (G) and the pyrimidines thymine (T) and cytosine (C). The single nucleotides polymerize connected by their phosphate groups, building up a complete DNA strand. See figure 3.1 for an overview of a DNA strand composed of the four different nucleotides of DNA. In the lower nucleobase the numbering of the carbon atoms of the sugar is depicted. Very important are the positions 3′ and 5′ which will from now on be used as a descriptor of the two possible ends of a DNA strand.

In the course of this work only double-stranded DNA will be investigated. The double-strands are formed by two single strands where all nucleobases form hydrogen bonds with their specific partners. In every case, a purine nucleobase forms a base-pair with a pyrimidine nucleobase. This way, there are two possible base-pairs from which a DNA double strand can be built up: Adenine-Thymine and Guanine-Cytosine. The former can form two hydrogen bonds, while the latter can form three hydrogen bonds and therefore is supposed to be more stable. Figure

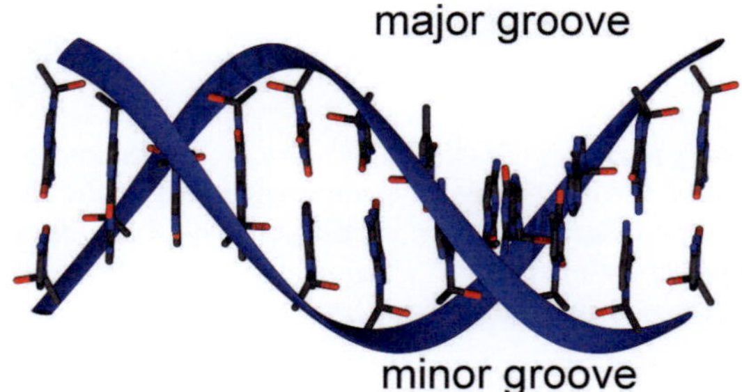

Figure 3.2: *Possible base-pairs in a DNA double-strand. Adenine-Thymine on the left side, Guanine-Cytosine on the right side. Hydrogen bonds are shown in red.*

Figure 3.3: *Typical form of a B-like DNA strand. The backbone is shown as a cartoon like representation in blue. The nucleobases are color coded by the elements. Major and minor grooves are recognizable here.*

3.2 shows the two possible combinations of nucleobases with the formed hydrogen bonds.

The DNA double-helix can adopt several conformations. Widely known are the A- and B-DNA conformation, which are both right-handed helices and most often occur in living cells. Despite there are more conformations known, like the left-handed Z-DNA for example, only A- and B-like DNA species are investigated in the course of this thesis. Besides several environmental parameters the DNA conformation also depends upon the sequence of the nucleobases[105].

A commonly described feature of the DNA double-helices are the grooves, which are formed around the helix. In the A- and B-like DNA one can distinguish between the major and the minor groove. Figure 3.3 shows a B-like DNA strand where these grooves are formed. In nature these grooves provide binding sites for proteins or other molecules involved in biological processes.

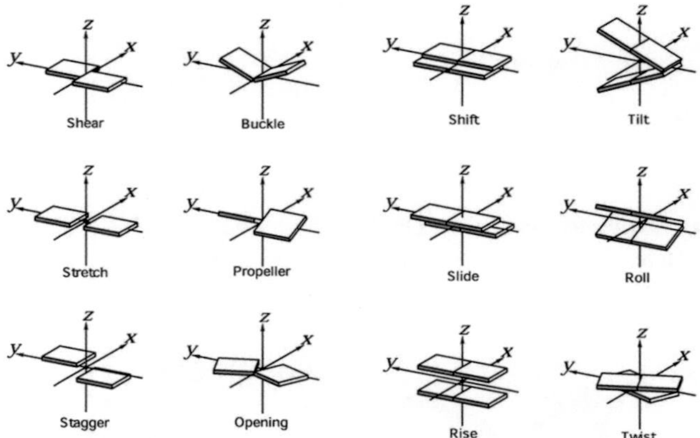

Figure 3.4: *Helical parameters of DNA double-helices. Six parameters on the left side represent the intra-base-pair parameters which depict the position of the two nucleobases which form a base-pair. The six parameters on the right side depict the inter-base-pair parameters which shows the position of two stacked base-pairs. Taken from [107]*

Helical Parameters

The structure of a DNA double helix can be described in terms of so-called helical parameters. The nomenclature and definition of these was specified by Dickerson et al. in 1989[106]. Until then no universal scheme to calculate these parameters was available and the nomenclature was inconsistent. The defined helical parameters describe the position of the nucleobases, while the structure of the backbone is not taken into account. Two groups of helical parameters are defined. First, the intra-base-pair parameters, which describe the deviation from a planar arrangement within a pair of nucleobases. Second, the inter-base-pair parameters which describe the orientation of two base-pairs stacked on top of each other. Figure 3.4 gives a graphical illustration of these 12 helical parameters. The helical parameters in this work were calculated with the 3DNA program.[107]

3.1.1   Investigated DNA Sequences

In the first part of this work the general features of the DNA double helix are investigated. The choice of the sequence is a non-trivial task as the humidity-dependent properties and behavior under tension may depend upon the sequence

Table 3.1: *The studied DNA fragments of the first part of this work. Shown are the descriptors used throughout this work and the nucleobase sequence of the well-matched dsDNA oligomers.*

	Homogeneous sequences
A5	5'-AAAAA-3'
	3'-TTTTT-5'
A9	5'-AAAAAAAAA-3'
	3'-TTTTTTTTT-5'
A13	5'-AAAAAAAAAAAAA-3'
	3'-TTTTTTTTTTTTT-5'
G5	5'-GGGGG-3'
	3'-CCCCC-5'
G9	5'-GGGGGGGGG-3'
	3'-CCCCCCCCC-5'
G13	5'-GGGGGGGGGGGGG-3'
	3'-CCCCCCCCCCCCC-5'

strongly. Both polyA and polyG DNA probably exhibit some special properties that are not found with other (irregular) sequences though. Yet, they are undoubtedly very different, and this provides a chance to see qualitatively what the sequence-related effects in such short oligonucleotides are.

Therefore, subject of the first study (chapter 4) are poly-adenine and poly-guanine double-stranded helical DNA species, containing five, nine and 13 base pairs; every species was capped with two GC base pairs on each end. Note that 13 base pairs correspond to slightly more than one complete helical turn in the B-DNA conformation. So, there are six dsDNA species in the first part, see table 3.1. Note that the polyA DNA species are unlikely to be thermodynamically stable under normal conditions. However, such A-tracts may occur as fragments within longer dsDNA, and the capping GC base pairs act as stabilizers of the double-stranded structure here.

In the proceeding chapters, a new set of DNA sequences will be introduced. As the general features of the DNA structure under mechanical stress and microhydration

Table 3.2: *The studied DNA fragments for charge transfer simulations. Shown are the two-letter symbols used throughout this work and the nucleobase sequence of the well-matched dsDNA oligomers. Purine nucleobases (G and A) act as charge carrying sites. (DD, Dickerson's dodecamer adopted from PDB ID 1BNA).[108]*

purines on one strand		purines distributed	
GG	5'-GGGGGGG-3'	GT	5'-GTGTGTG-3'
	3'-CCCCCCC-5'		3'-CACACAC-5'
AA	5'-AAAAAAA-3'	GC	5'-GCGCGCG-3'
	3'-TTTTTTT-5'		3'-CGCGCGC-5'
GA	5'-GAGAGAG-3'	AT	5'-ATATATA-3'
	3'-CTCTCTC-5'		3'-TATATAT-5'
		DD	5'-CGAATTCG-3'
			3'-GCTTAAGC-5'

are investigated and the simulation setup is generated, more sophisticated DNA sequences were needed for the calculation of CT in DNA. Therefore, seven different dsDNA oligomers were studied, three oligomers with purine nucleobases located exclusively on one strand in the central fragment, and four oligomers with purines distributed among both strands. The seven innermost base-pairs were taken into account (eight for the Dickerson dodecamer (DD) for later analysis, to avoid analyzing effects which only occur at the ends of the DNA strands. These fragments and their descriptors used throughout this work are presented in Tab. 3.2.

3.2 MD Simulation of DNA

The 3D structural model of each oligonucleotide was build in an idealized B-DNA conformation with the na-server.[109] The molecular system then was enclosed in a periodic rectangular box keeping a minimum distance from the walls of the box of 1 nm. For the DNA stretching simulations this minimal box was then extended by additional 5 nm along the z-axis, parallel to the DNA helical axis, to accommodate a stretched conformation of the DNA molecule. The box was filled with TIP3P water molecules[110] and the density of liquid water was reproduced. An appropriate number of water molecules, depending upon the length of DNA, were replaced with sodium counterions to neutralize the negatively charged DNA backbones.

Classical MD simulations were performed with the parm99/BSC0 force-field,[111–113] and the parameterization by Åqvist was applied to describe the Na^+ counterions.[114] The topologies were created with AmberTools[115] and converted to the Gromacs format with Ambconv.[116] All simulations were performed with a time step of 2 fs with LINCS[117] to constrain the bonds involving hydrogen to their reference lengths. Pilot calculations, performed on nucleobases in the gas phase as well as on a solvated oligonucleotides, showed that these constraints affect neither the ionization potentials of the nucleobases nor the electronic couplings between them. The Lennard-Jones interactions were cut off at 1 nm, and the electrostatics were treated with the particle–mesh Ewald method [52].

The prepared systems were equilibrated in a multi-step procedure. First of all, a steepest-descents minimization was performed for 100 steps to remove possible bad contacts. Subsequently, the initial velocities for all atoms were randomly drawn from a Maxwell–Boltzmann distribution at 10 K. The water was heated up first in an NVT simulation of 20 ps length. Here the DNA was weakly coupled to a bath at 10 K and the solvent coupled to a separate bath at 300 K. The Berendsen thermostat was used for this purpose.[57] After that, another NVT simulation of 20 ps length was used to bring the entire system to 300 K, with a single heat bath. Finally, an NPT simulation at 300 K and 1 bar was performed over the interval of 0.5 ns. The Nosé–Hoover extended-ensemble thermostat with a characteristic time of 0.5 ps was used to keep the temperature of 300 K.[58, 59] In NPT simulations, the Parrinello–Rahman barostat with characteristic time of 0.5 ps and a reference pressure of 1 bar was used.[60] The consecutive production simulations were run with the same parameters as the last equilibration step. The Gromacs package (versions 4.0 and 4.5)[118, 119] was used for all simulations.

3.3 DNA under Mechanical Stress

Each DNA species was stretched by pulling its 3'-ends in opposite directions. This setup was inspired by the experiments where the 3'-ends of dsDNA were contacted to the electrodes. To do so, an additional force along the z-axis of the simulation

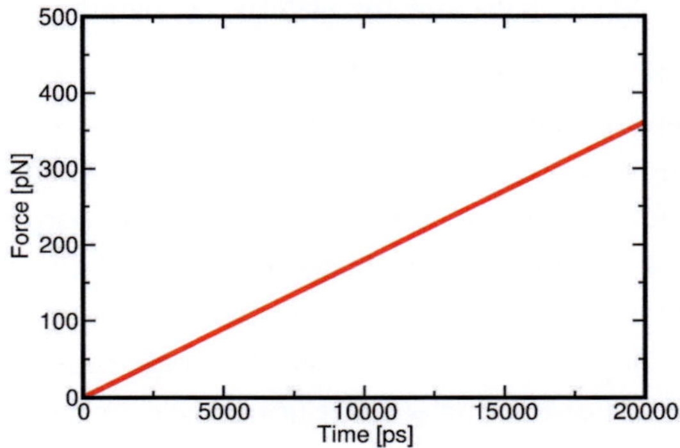

Figure 3.5: *Example of the increasing pulling force during an MD simulation of 20 ns of the DNA oligomer A_5.*

box, parallel with the helical axis of the oligomer, was applied to the O3′ atom of the 3′-terminal nucleotide in each DNA strand. Evidently, the approach involving such 'artificial' stretching of the DNA molecule does not involve the effect of variable distance of the DNA to the electrodes. In other words, it is assumed here that the conformational changes of the DNA molecules take place before the coupling of DNA to the electrodes is affected considerably. The force on the O3′ atoms was increased with different rates along the simulation until the DNA strands separated. See figure 3.5 for an example of the linearly increased pulling force during an MD simulation of 20 ns.

Note that the used loading rates are several orders of magnitude larger than that applied in typical experiments. The likely consequences of this fact is a more distinct irreversibility of the stretching process. This will be discussed further on in chapter 4.

3.4 Microhydrated DNA

To prepare the microhydrated systems a part of the water molecules was removed from the equilibrated fully hydrated systems. Pilot simulations showed that a hydrated backbone is best to support the helical character of the DNA molecule.

Table 3.3: Number of water molecules in the microhydrated systems.

system	waters per		A_5 A_9 A_{13}			ratio to full
	phosphate	base pair	G_5	G_9	G_{13}	hydration
Dry1	24	ca. 43	384	576	768	1/10
Dry2	12	ca. 21	192	288	384	1/20
Dry3	6	ca. 11	96	144	192	1/40
Dry4	3	ca. 5	48	72	96	1/80

To keep the water molecules nearest to the backbone, the distance between the O-atoms of the water molecules and the P-atoms of the backbone was measured and the nearest water molecules were kept. This way, four different microhydration states were created. These obtained systems contained from 24 down to 3 water molecules per phosphate. See Table 3.3 for an overview of the systems naming and the amount of water left in the simulation box.

This corresponded to the removal of ca. 9/10, 19/20, 39/40 and 79/80 of water from the fully hydrated system, respectively. The appearance of the DNA hydration shells constructed in this way can be inferred from 3.6.

These microhydrated systems *in vacuo* were simulated in an NVT ensemble. Pilot simulations showed some unwanted effects when using cut-offs to deal with the electrostatics in the simulation of these clusters. Since Gromacs is not able to calculate Ewald in multi-threaded environment, periodic boundary conditions were applied to make use of the particle-mesh Ewald method. To avoid undesired boundary effects the box size was increased here to $(11,5 \text{ nm})^3$. To obtain stable MD simulations of these molecular clusters, it was necessary to decrease the time step to 1 fs. Again, the parm99/BSC0 force field was used, in spite of the fact that it had been developed for simulations in condensed phase. In the parameterization, the atomic charges are overestimated by 15–20 % systematically, compensating for the missing description of electronic polarization effects. This may be regarded as an error. However, the aim is to characterize effects of robust nature, and so this is most likely not an issue. Indeed, it was shown that the interactions between nucleic acid building blocks are described well.[120, 121]

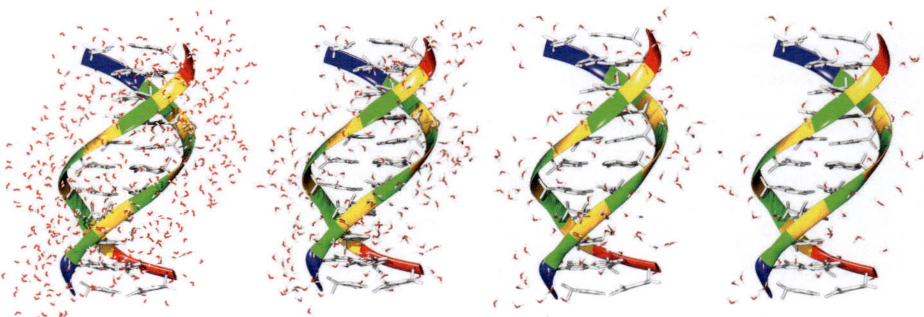

Figure 3.6: *Studied microhydrated systems – A_{13} with varied number of water molecules.*

An additional constant external force was applied to the microhydrated DNA oligomers, emulating the experimental setup where the oligomer is held between two electrodes. A series of trial simulations with varied magnitude of the external force was performed to estimate the force that would support a double-stranded structure of DNA. This force was applied as an harmonic constraint with a force constant of 1000 kcal/(mol nm) between the O3'-atoms of the outermost nucleotides.

Also studied were the systems Dry1 and Dry2 with a smaller number of counterions than required for neutralization. The clusters were stripped randomly of 1/4, 1/2, 3/4 as well as off all ions, respectively. MD simulations were performed for these negatively charged systems with the same setup as described above.

The simulations of microhydrated DNA with an external pulling force of increasing magnitude were performed in the same fashion as those of the fully hydrated DNA. A single loading rate of 10 pN/ns was applied here.

CHAPTER 4

DNA Under Experimental Conditions

Reproduced in part with permission from
M. Wolter, M. Elstner and T. Kubař,
J. Phys. Chem. A, 2011, 115 (41), 11238–11247.
Copyright 2011, American Chemical Society.

MD simulations of DNA, in course of charge transfer simulations, are usually not performed under conditions resembling the experimental setup of electric conductivity experiments. The DNA strands are connected to electrodes on both sides, causing mechanical forces onto the double helical structure. Moreover, the DNA strands are dried in some fashion to ensure the measurements are not influenced by the conductivity of water.

Therefore the first task for the simulation of this experimental setup is to investigate the structure of double-stranded DNA under mechanical stress in complex with varied amount of water and counterions. To get a first insight into the structure of DNA and the changes under mechanical stress, MD simulations are performed with homogeneous sequences. Apart from the mechanical stress, the influence of the environment on the structure and stability of the DNA double-strand is investigated. Various amounts of water, as well as the counterions are seen to contribute to the stabilization of the DNA structure.

4.1   Free MD Simulations

In order to obtain reference values, the starting point will be a brief analysis of the structure of free, fully hydrated DNA. MD simulations of 20 ns in aqueous solution were performed for every DNA species studied, and the structure of each of them was observed stable throughout the simulation time. The length of each double strand was measured as the distance of the 3'-end O3' atoms, see table 4.1 for the mean values and standard deviations. The distributions of these lengths are shown in figure 4.1.

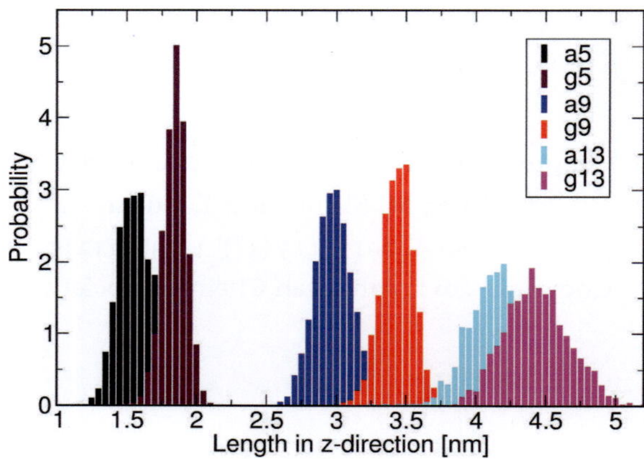

Figure 4.1: *Distribution of length of studied DNA species in aqueous solution.*

Table 4.1: *Length of the studied DNA species in aqueous solution in free MD simulations, measured as the distance of 3'-end O3' atoms (mean $\pm$ standard deviation).*

sequence	length [nm]	sequence	length [nm]
A_5	1.57 ± 0.13	G_5	1.84 ± 0.08
A_9	2.98 ± 0.14	G_9	3.44 ± 0.12
A_{13}	4.17 ± 0.22	G_{13}	4.44 ± 0.23

In comparison to the polyA oligomers, the polyG oligomers exhibit a larger end-to-end distance for all sequences with matching numbers of base pairs. This may seem surprising as the B-like polyA strands are actually longer then the A-like polyG. The reason for this is that the 3′–3′ distance does not correspond exactly to the length of the DNA molecules measured along the helical axis. Nevertheless, the 3′–3′ distance will be considered in this work because it is the coordinate that DNA will be pulled along, motivated by the single-molecule conductivity experiments. Interestingly, the fluctuation of this distance is smaller for polyG than for polyA with shorter DNA double strands of five and nine base pairs, while it is larger for G_{13} than for A_{13}. The absence of any obvious trends may be caused by the fact that the studied DNA oligomers are quite short, with only A_{13} and G_{13} constituting one complete helical turn.

Also, the helical parameters rise, slide and twist were evaluated for the studied DNA species and are presented in table 4.5. This table shows the mean values of the helical parameters of all considered base-pair steps. For an overview over all parameters for every single step, see tables A.1 and A.2 in the appendix. The smaller values of twist and the larger negative values of slide in case of polyG sequences indicate these DNA strands to assume an A-like conformation, which is in accordance with expectations. However, no clear difference between polyA and polyG was observed for rise.

4.2 The Structural Changes of DNA upon Stretching

A series of simulations was performed in which an additional external force was applied on the 3′-end O3′ atoms of the particular DNA species. To simulate the stretching within a tractable simulation time, the magnitude of the external force was increased gradually in time with the rate of 50 pN/ns (83 pN/ns for A_{13} and G_{13}). This way, the force at the end of an MD simulation of 20 ns reached a value of 1.00 nN (1.66 nN for A_{13} and G_{13}). This turned out to be sufficient to finally separate the strands in all of the studied DNA species. As a measure of length of the DNA oligomer, the distance of the 3′-end O3′ atoms was monitored along the simulation. See figure 4.2 for an example profile of this distance vs time or, equivalently, external force, and figure 4.3 for a set of representative snapshots.

The helical double-stranded structure of the A_9 oligomer is preserved up to ca. 200 pN. Then, the end-to-end distance increases by over two thirds of the initial length within a very short interval of time. This indicates the transition to a struc-

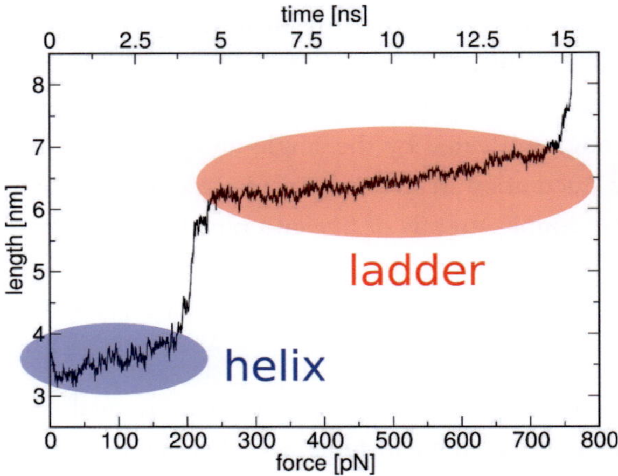

Figure 4.2: *Stretching of A$_9$ in aqueous solution – the elongation–force profile. The regions where helix and ladder structure appear are designated.*

ture that is still double-stranded but not helical any more. In this newly adopted structure, the strands are parallel to each other while maintaining the hydrogen bonding between the nucleobases. This structure corresponds to the proposed S-DNA,[21] and is designated as 'ladder' in the following. The ladder remains stable for a relatively long interval of time and increasing force. Finally, the DNA strands separate at ca. 750 pN. This sequence of events is observed for all studied DNA species, with the separation of strands occurring at a force that ranges from 450 pN (for A$_5$) to 1300 pN (for G$_{13}$).

Also, the monitoring of the DNA helical parameters along the stretching simulations captures both conformational transitions. The ladder structure exhibits a twist of around zero, compared to above 30 deg. for A-/B-DNA, and a slide of ca. +0.5 nm, compared to -0.15/0.00 nm for A-/B-DNA. See figure 4.4 for these data on A$_9$.

Upon the separation of DNA strands, the helical parameters lose their meaning whatsoever. But, once they are calculated, extremely large oscillations are observed.

Another useful tool is the analysis of the pattern of interstrand hydrogen bonding between the nucleobases. See figure 4.5 for an example pattern calculated for the oligomer A$_9$. When the ladder structure is torn, all hydrogen bonds vanish naturally. Interestingly, visible are also further irregularities in the hydrogen bonding

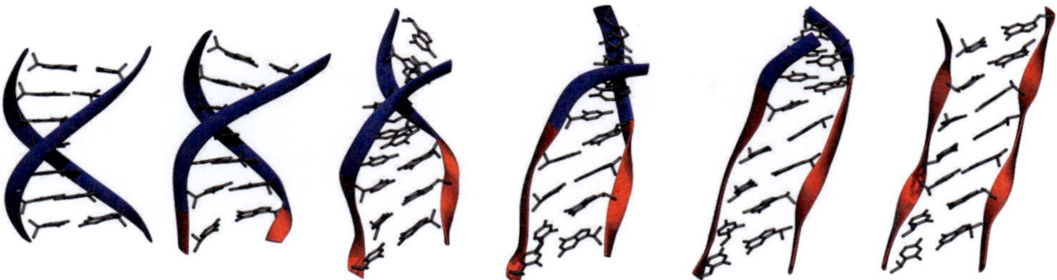

Figure 4.3: *The central seven base pairs of the DNA oligomer GG being stretched up to the additional extension of ca. 25 % of the free length by pulling at the 3'-ends. The backbones are depicted as ribbons, with color coding the conformation: blue – helical, B-like dsDNA; red – ladder-like S-DNA. The helical dsDNA structure passes to a ladder-like one, gradually.*

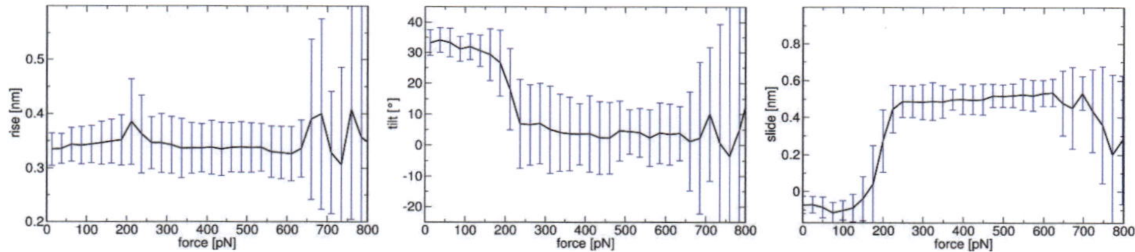

Figure 4.4: *Helical parameters in a stretching simulation of A$_9$.*

patterns.. Shortly after the helix–ladder transition (at 250–300 pN), the hydrogen bonding in a few central base pairs is disturbed for a prolonged period of time. Partially, these defects pertain for the entire rest of the simulation, showing that the transition to a ladder structure need not be perfect. The analysis of hydrogen bonding during stretching for all studied DNA oligomers is presented in the appendix (figure A.1).

An important issue of such simulations is the dependence of observed structural changes on the rate of stretching. Note, that the stretching is several orders of magnitude slower in the experiments. To obtain a rough idea on the effect of rate on the stretching profile, further simulations were performed. The external force was increased gradually with the rate of 10 pN/ns and even 2 pN/ns (for A$_9$ only), instead of the previously used rate of 50 or 83 pN/ns. The comparison of stretching profiles at three different loading rates is presented in figure 4.6.

Conformational transitions occur at smaller force when the force increases more slowly, even though the emerging conformations are the same. Instead of at

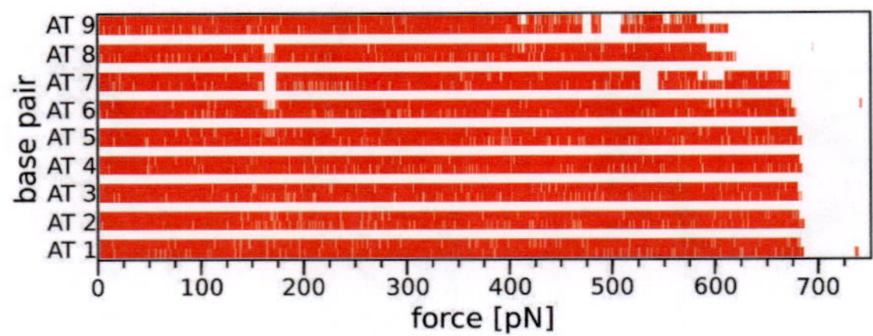

Figure 4.5: *The Watson–Crick hydrogen-bonding pattern in the A_9 oligomer during stretching. Red color denotes an existing hydrogen bond at the corresponding magnitude of force. Note the temporary break of hydrogen bonds in several base pairs during the overstretching transition at 170 pN.*

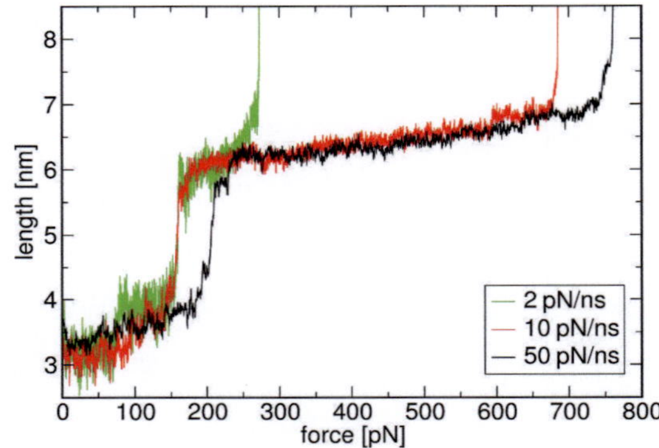

Figure 4.6: *Profiles of distance vs external force for the stretching of A_9 with different loading rates.*

200 pN, the formation of the ladder is observed at 150 pN at the smaller loading rates. Even more markedly, the separation of strands follows at quite large force of 750 and 700 pN at the larger loading rates of 50 and 10 pN/ns, respectively, but already at 270 pN at the slowest loading rate of 2 pN/ns. This fact hints at the non-equilibrium character of the process being simulated. The question of irreversibility was addressed in more detail by means of removal of the external force, and the observations are presented in the next section.

The dependence of the length on the external force, or simulation time is essentially identical with all the DNA species studied. See figure 4.7 for the extension–force profiles obtained at 10 pN/ns.

The end-to-end distance is expressed as a relative value with respect to the average length in the free simulation performed before. In a first interval after the

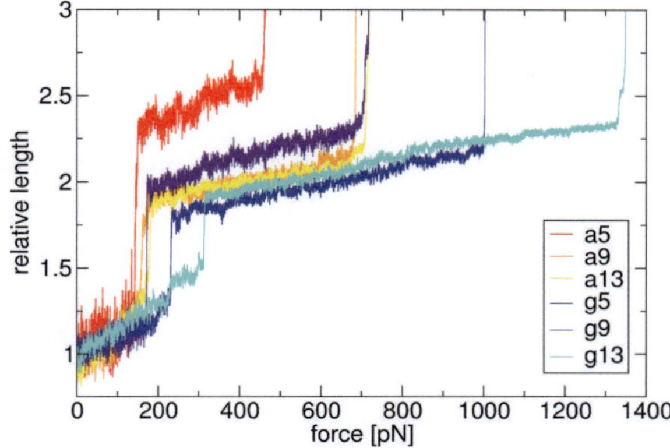

Figure 4.7: *The stretching profiles (relative extension vs. applied external force) for all studied DNA oligomers, at a stretching rate of 10 pN/ns.*

introduction of the external force, all DNA oligomers exhibit a roughly linearly increasing length, behaving nearly like harmonic springs. Then, the transition to a ladder structure sets in at some point at an external force of between 150 and 300 pN, depending on the length and composition of the DNA oligomer. The ladder structure, with an end-to-end distance about twice as large as that of free helical double-stranded DNA, remains stable for another extended interval of time while the force further increases. Finally the forces are high enough to reach up to a point where the DNA strands separate entirely. Obviously, the longer DNA oligomers undergo the helix-to-ladder transition as well as the eventual separation of strands at a larger external force. The force needed to induce a transition increases in the series $A_5 > A_9 > A_{13}$ and $G_5 > G_9 > G_{13}$. Also, the guanine-containing oligomers require generally a larger force to assume the ladder structure as well as to tear apart. The situation would become more complex with mixed DNA sequences surely.

The stretching simulations were performed with a setup very similar to that used in earlier theoretical studies. [19, 20, 22, 33] The obtained results are in accordance with these respective data. Specifically, the pulling rates were in the same range, the onset of the ladder-like overstretched structure was observed at similar forces, and the overstretched DNA itself possesses the same structural features. In the following, these stretching profiles will serve as a reference to which the behavior of microhydrated DNA during stretching will be related.

4.3 Irreversibility of DNA Stretching in Simulations

The external force applied in the simulations increases several orders of magnitude faster than in the relevant experiments. In addition, the observed different forces needed for the transition when different loading rates are applied, provides a hint at the likely irreversibility of the simulated process. So, the important question here is what happens with the DNA oligomer when the external force is removed – does the helical structure recover? To investigate this point, the external force was removed at several selected points during each simulation, which was followed by a free simulation of 20 ns. Also, an alternative protocol was used with the force being decreased gradually by 10 pN/ns down to zero instead of its abrupt removal. The points of force removal or reversal were selected as follows:

- A – helix just before the transition to ladder

- B – during the helix–ladder transition

- C – freshly established ladder

- D – ladder just before the final collapse of the double strand

The time and force at these points in each simulation is provided in table 4.2.

Table 4.2: Time (t, in ns) and force (F, in pN) at the point in a stretching simulation that reverse simulations A, B, C and D were started, for the DNA species studied.

Sim.	A_5		A_9		A_{13}		G_5		G_9		G_{13}	
	t	F	t	F	t	F	t	F	t	F	t	F
A	13.5	81	12.0	72	16.0	96	16.5	99	22.0	132	30.0	180
B	15.0	90	16.0	96	18.0	108	17.3	104	23.3	140	31.5	189
C	16.5	99	21.0	126	20.0	120	18.0	108	24.0	144	33.0	198
D	42.0	252	60.0	360	65.0	390	65.0	390	95.0	570	132.0	792

The typical distance/force profile in both kinds of simulations is presented in figure 4.8, and the character of the observed final structures is summarized in table 4.3.

When the external force was switched off in points before or during the helix–ladder transition (A, B), the helix structure was always preserved/recovered. This

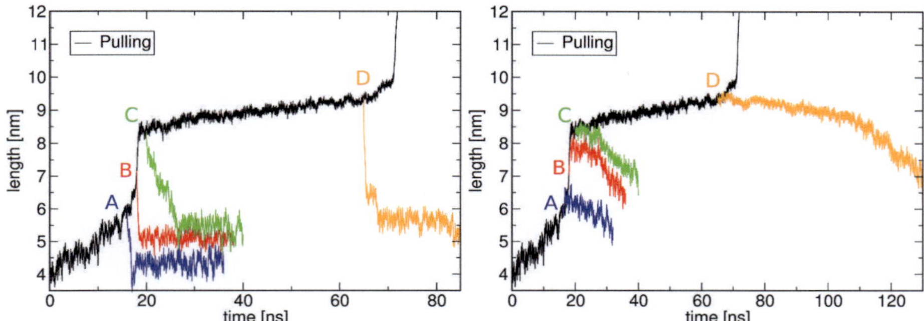

Figure 4.8: *Distance/force profile in the stretching/relaxing simulations of A_{13}. Black – stretching (increasing force), colored – subsequent relaxation from the points A, B, C and D; left – external force removed abruptly, right – force decreased to zero gradually.*

Table 4.3: *Final structures after the stretching/relaxing simulations.*

	force removed abruptly					force decreased gradually			
	from A	from B	from C	from D		from A	from B	from C	from D
A_5	helix	helix	helix	ladder	A_5	ladder	ladder	ladder	ladder
A_9	helix	helix	ladder	helix	A_9	helix	ladder	ladder	separation
A_{13}	helix	helix	helix	helix	A_{13}	helix	ladder	ladder	ladder
G_5	helix	helix	ladder	ladder	G_5	ladder	ladder	ladder	separation
G_9	helix	helix	helix	ladder	G_9	ladder	ladder	ladder	separation
G_{13}	helix	helix	helix	helix	G_{13}	ladder	ladder	ladder	separation

was not always the case when the force was removed already in the ladder state of the DNA oligomer. In some simulations, the ladder was preserved even in the free simulation whereas the other simulated DNA species returned to a helical conformation.

The situation changes dramatically when passing to the simulation with a gradual decrease of the force. Already when the external force was being switched off from the midpoint of the helix–ladder transition (B), the helical conformation does not recover within the simulation time. This hysteresis emphasizes the irreversible character of DNA stretching in the performed simulations. Its even more distinct demonstration is the separation of DNA strands observed in four of the six simulations after the reversal of external force in point D.

4.4   Effects of Low Hydration

To model the low hydration of DNA, most of the water molecules were removed from the simulation box. Since the exact amount of water remaining in the cluster is not known in the single-molecule experiments, four different levels of hydration were probed in this work. Four sets of DNA–water clusters were prepared for every DNA sequence, containing 1/10 (Dry1), 1/20 (Dry2), 1/40 (Dry3) and 1/80 (Dry4) of the original number of water molecules, respectively, see table 3.3. Free MD simulation was then commenced for 10 ns, and the structure of the DNA oligomers was monitored.

Table 4.4: Appearance of the microhydrated systems after an MD simulation of 10 ns with no external force. Helical – structure close to that with full hydration, with no or minimal curvature; bent – structure that lost the linear shape yet with preserved helical character; distorted – structure that lost the helical character but with observable hydrogen-bonding pattern; coiled – severely deformed structure with corrupted hydrogen bonding of nucleobases.

sequence	Dry 1	Dry 2	Dry 3	Dry 4
A_5	helical	bent	bent	coiled
A_9	helical	bent	distorted	coiled
A_{13}	bent	bent	distorted	coiled
G_5	helical	bent	distorted	coiled
G_9	helical	bent	distorted	coiled
G_{13}	helical	bent	distorted	coiled

The structures of all the studied systems are summarized in table 4.4. The helical structure of DNA in the Dry1 systems is generally retained, however with some defects. The A-like favoring oligomer G_{13} forms a structure with an even larger diameter than A-DNA but shorter, a sort of a barrel structure, see figure 4.9.

The species A_{13}, on the other hand, bends toward the major groove so that the ends of the double strand come closer to each other. The major groove is extremely narrow in the bend, and there are several Na^+ ions in the groove, preventing unfavorable contacts of the phosphates in the backbones, see figure 4.10.

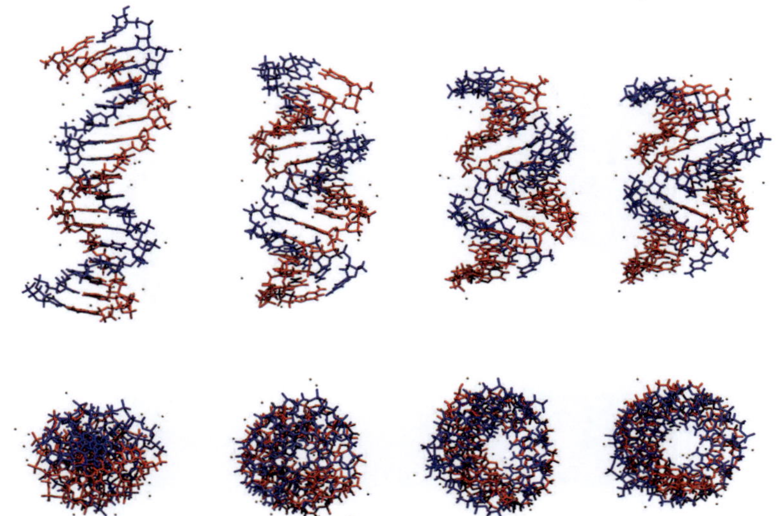

Figure 4.9: *The microhydrated DNA species* G_{13} *(Dry1) adopts a more compact (barrel-like) structure in an MD simulation with no stabilizing external force. (Side as well as top view; surrounding water molecules not shown for clarity.)*

The difference between polyA and polyG can be explained by the larger hydration of polyA and its known inability to assume an A-like structure. Indications of these effects are generally visible in the shorter DNA oligomers, too. The DNA oligomers lose the helical character in the Dry3 systems, and they collapse to disordered compact structure in Dry4, where even the hydrogen bonding of nucleobases disappears. All of these quite different conformational transitions may be seen as a pursuit to assume a more compact conformation, under given conditions.

The observed irregularities may be overcome by the application of a moderate external pulling force acting against the tendency of the insufficiently hydrated DNA double strand to bend. For this reason as well as to emulate the setup of conductivity experiments, further simulations were performed with an additional external force applied to the O3' atoms on the 3'-ends of each DNA strand. The magnitude of this force was chosen in such a way that the DNA would retain straight helical double-stranded structure, which turned out to be 50 kJ/mol·nm = 83 pN for Dry1 and 100 kJ/mol·nm = 166 pN for Dry2. Selected helical parameters were evaluated along these simulations, see table 4.5. In the Dry3 and Dry4 systems, no helical

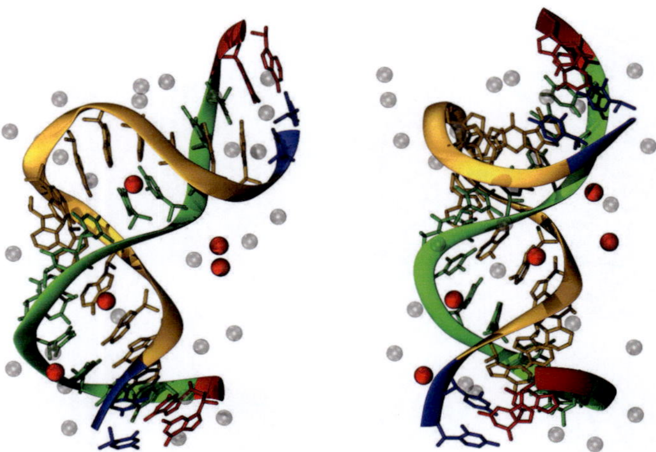

Figure 4.10: The structure of microhydrated DNA species A_{13} collapses in an MD simulation with no stabilizing external force. (Surrounding water molecules not shown for clarity.)

pattern was observed in the DNA structure any longer, and the helical parameters were not evaluated.

The largest studied amount of water (Dry1) was sufficient to support nearly-canonical structure of all DNA oligomers when the stabilizing pulling force was applied. All of the polyA species stayed in B-like conformation in spite of the low hydration, illustrating that it is difficult to convert polyA to A-DNA, which would generally be preferred under low hydration. Slightly unexpected was the observation of helical parameters taking rather B-like values for G_5. Although, this may not be surprising for a DNA species much shorter than a single helical turn, and it is also not quite clear if the applied force field describes a B-to-A conversion with sufficient accuracy. A further reduction of the amount of water (Dry2) leads to structures that were too irregular to be assigned to either A- or B-family. Also, these structures fluctuate notably, which is reflected by the large standard deviation of helical parameters, see table 4.5.

The degradation of structure can be monitored also by the analysis of the hydrogen-bonding pattern between the DNA strands. An example in figure 4.11 shows the hydrogen-bond existence charts from MD simulations of 10 ns of the microhydrated A_{13} species (Dry1 and Dry2), still with the 3'-ends pulled with an external force. Evidently, all of the hydrogen bonds are preserved in the Dry1 simulation,

Table 4.5: *Helical parameters of the studied DNA species in microhydrated environment (Dry1 and Dry2) in comparison with the values obtained for DNA in aqueous solution. Presented are mean values and standard deviations evaluated along MD simulations and averaged over the considered base-pair steps. For the data obtained for the individual base-pair steps, see tables A.1 and A.2 in the appendix*

sequence	rise [nm]			twist [deg.]			slide [nm]		
A-DNA	0.320			32.0			−0.15		
B-DNA	0.340			36.0			0.00		
	complete	Dry1	Dry2	complete	Dry1	Dry2	complete	Dry1	Dry2
A_5	0.335 ± 0.030	0.334 ± 0.028	0.303 ± 0.083	33.6 ± 5.1	36.6 ± 4.9	25.0 ± 25.0	-0.061 ± 0.057	0.004 ± 0.040	0.078 ± 0.093
A_9	0.335 ± 0.031	0.328 ± 0.028	0.337 ± 0.055	33.7 ± 5.0	36.0 ± 4.8	41.8 ± 13.7	-0.065 ± 0.056	0.000 ± 0.043	0.115 ± 0.093
A_{13}	0.334 ± 0.031	0.326 ± 0.027	0.323 ± 0.046	33.0 ± 4.9	37.3 ± 4.8	41.1 ± 13.2	-0.074 ± 0.057	0.000 ± 0.044	0.052 ± 0.082
G_5	0.340 ± 0.033	0.336 ± 0.032	0.330 ± 0.032	29.3 ± 4.3	34.8 ± 4.9	39.0 ± 6.2	-0.142 ± 0.058	-0.027 ± 0.064	-0.011 ± 0.077
G_9	0.339 ± 0.032	0.328 ± 0.026	0.313 ± 0.064	29.0 ± 4.2	29.7 ± 4.3	34.0 ± 13.1	-0.143 ± 0.056	-0.142 ± 0.053	-0.013 ± 0.082
G_{13}	0.341 ± 0.033	0.329 ± 0.026	0.332 ± 0.042	29.5 ± 4.3	29.2 ± 4.5	35.3 ± 5.7	-0.136 ± 0.060	-0.142 ± 0.048	0.034 ± 0.069

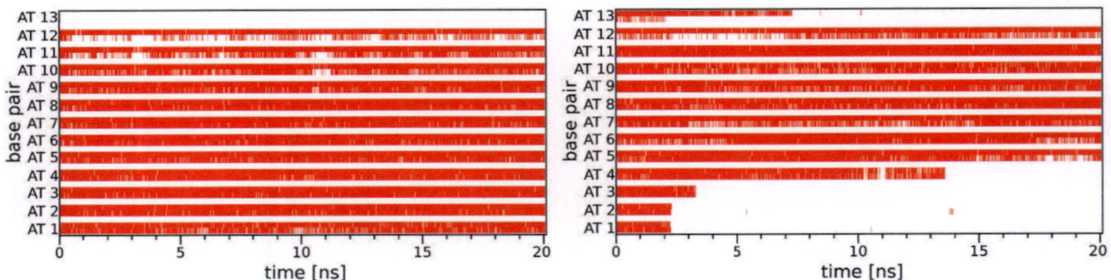

Figure 4.11: *Watson–Crick hydrogen-bonding pattern of the microhydrated species A_{13} in simulation with stabilizing external force – Dry1 (top) and Dry2 (bottom). Red – hydrogen bond present, white – no hydrogen bond.*

and there are several defects in the Dry2 case. The hydrogen-bonding pattern in the shorter DNA oligomers is more vulnerable and is destroyed almost entirely within a few nanoseconds in the Dry2 systems. See figure A.2 in the appendix for a complete overview of the hydrogen pattern of the different sequences.

4.5 Effects of Decreased Ion Content

Another important component of the molecular system is the counterions, Na^+ in this case. In an extreme case of completely dehydrated DNA with no ions, the helix would unwind, and the strands would separate readily due to the repulsive electrostatic interaction of the negatively-charged backbones. Note that the phosphate groups in the backbones retain their negative charges throughout this work. While a partial protonation of phosphates was considered in a previous study, where DNA under harsh conditions (electrospray) was modeled,[122], this work aims at the description of DNA in the milder setup of single-molecule experiments where the solvent is removed partially with a stream of gas. The most straightforward assumption is that of unchanged charge state of phosphates, which may be neutralized by counterions remaining in the hydration shell. Then, the question is whether the number of counterions necessary for neutralization would remain in the cluster or rather be carried away by the stream of gas together with the solvent. It is not possible to provide an answer based on the simulations in this work, yet what can be done is to investigate if and how the stability of DNA structure changes when the counterions are (partially) removed, at a given degree of hydration. To do so, the systems Dry1 and Dry2 from the previous section (with 1/10 and 1/20 of the original amount of water, respectively) were stripped further of 1/4, 1/2 and 3/4 of the original number of sodium ions as well as of all of them. MD simulations were then performed with no external force. The observed structural changes are described in table 4.6.

Table 4.6: Appearance of the Dry1 complexes after MD with a decreased number of sodium ions. B-like – long helical structure with a small diameter, A-like – shorter helical structure with a large diameter, ladder – parallel linear double-stranded structure.

	3/4 ions	1/2 ions	1/4 ions	0 ions
A_5	B-like	B-like	ladder	separation
A_9	B-like	B-like	ladder	separation
A_{13}	B-like	B-like	ladder	separation
G_5	A-like	B-like	ladder	ladder
G_9	A-like	B-like	ladder	separation
G_{13}	A-like	B-like	ladder	separation

In the complexes with more water (Dry1), the helical double-stranded structure of DNA was preserved even upon the removal of half of the ions. Interestingly, the helical structure of the studied DNA species is more stable with a slightly smaller number of ions than necessary for neutralization. The transition to a ladder structure occurred only after the removal of further 1/4 of the original number of ions. These structures resemble those reported in previous simulations of DNA with similar charge state[122] closely. Even though the simulated systems were very different otherwise (no water whatsoever, no counterions and phosphate groups neutralized partially). Eventually, the DNA strands separated when all of the ions were removed.

The structure of more poorly hydrated DNA oligomers (Dry2) was more prone to perturbations caused by the removal of ions, as the ladder structure was observed already with 1/2 of the amount of ions necessary for neutralization. Nevertheless, the separation of DNA strands took place only after the complete removal of counterions. Another interesting effect is that the polyG oligomers passed to a B-like conformation when 1/2 of the ions were removed. This may be interpreted in terms of the significance of counterions to maintain the A-like DNA conformation.

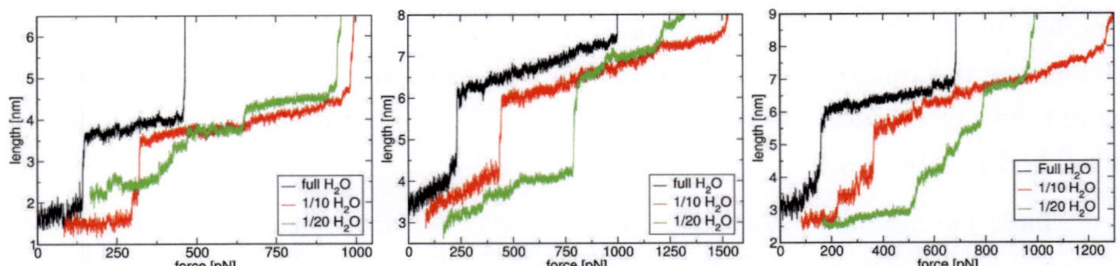

Figure 4.12: *Profiles of distance vs force for the stretching of microhydrated DNA – A_5 (left), A_9 (middle) and G_9 (right).*

Obviously and not surprisingly at all, the structure of helical double-stranded DNA was shown to be stabilized by both water and counterions, with the absence of one or the other bringing about a more or less severe perturbation of the structure. However, the helical structure of a microhydrated dsDNA species may be stabilized by the removal of a certain part of the counterions.

4.6 Effect of Water and Ions on the Stretching Profile of DNA

Finally, MD simulations with a gradually increasing pulling force were performed for the clusters Dry1 and Dry2. The simulations were started with a nonzero initial external force of 83 pN for Dry1 and 166 pN for Dry2 to maintain a helical double-stranded structure. Then, the force was increased during the simulation with the rate of 10 pN/ns.

Examples of stretching profiles of microhydrated DNA oligomers are presented in figure 4.12. See also figure A.3 for selected helical parameters during the stretching of microhydrated G_9. The microhydrated G_9 remains in the helical structure until the external force reaches 400 pN with Dry1 and even 800 pN with Dry2. Remember, this is the fourfold of the force needed to transform this DNA oligomer to a ladder structure in aqueous solution. This effect was observed for all DNA oligomers studied, see also table 4.7 for the forces at which overstretching transition occurs.

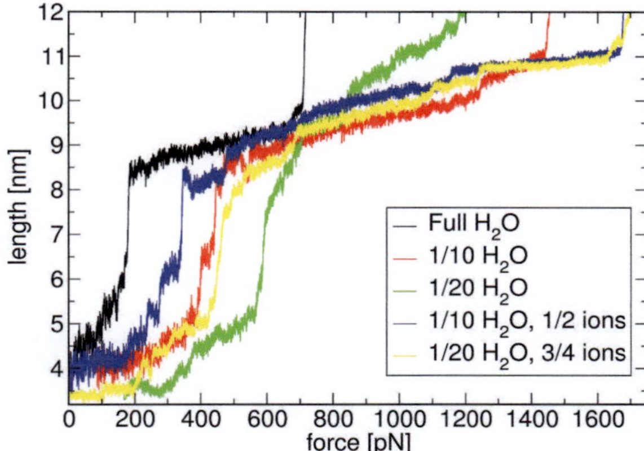

Figure 4.13: *Stretching profiles of DNA A$_{13}$ in a microhydrated environment with a smaller number of ions.*

It can be interpreted in two quite independent ways:

1. the helical structure of microhydrated double-stranded DNA is provided support by the external stretching force of moderate magnitude, acting against the bending of the oligomer, as conceived before.

2. this structure can resist a larger mechanical stress than the DNA in water can.

However, the situation is not quite unambiguous, as can be seen on the example of the short oligomer A$_5$, see figure 4.12 (left). Here, the double-stranded structure is still more stable in the Dry1 hydration shell than in bulk solution, but the Dry2 hydration is already insufficient to support any kind of helical pattern. Consequently, no clear conformational transition can be seen in the stretching profile, and the length of the double strand increases gradually up to a point where the strands separate. This kind of behavior was observed for this quite short DNA oligomer only. Another anomaly could be thought of in case of longer oligomers, where some parts of the double strand would remain in the helical conformation longer (i.e. up to a larger external force) than others that would pass to a ladder structure more readily. An example of a stretching profile that corresponds with such a hypothesis is that of the A$_9$ species with hydration shell Dry2 (see figure 4.12 (right)): The increase of distance vs force consists of several steps that would represent the successive transitions of different parts of the double strand to the ladder conformation.

Table 4.7: *Force at which the overstretching transition as well as strand separation (melting) occurred. Data from simulations of fully hydrated systems (with two different loading rates) as well as for the systems Dry1, Dry2 and Dry1 with a decreased number of sodium ions. All figures in pN; '—' denotes cases where no clear transition was observed. [a] The loading rate was 50 pN/ns for A_5, A_9, G_5 and G_9 while it was 83 pN/ns for A_{13} and G_{13}.*

	transition	fully hydrated		Dry1	Dry2	Dry1 with 1/2 ions
		50/83 pN/ns[a]	10 pN/ns			
A_5	overstretching at	187	142	288	—	184
	separation at	508	458	901	941	860
A_9	overstretching at	199	154	361	—	—
	separation at	746	678	1229	1109	946
A_{13}	overstretching at	293	173	395	580	335
	separation at	1014	708	1365	1331	1674
G_5	overstretching at	172	171	301	309	121
	separation at	713	704	932	1112	864
G_9	overstretching at	421	231	432	788	—
	separation at	—	1001	1444	1451	895
G_{13}	overstretching at	397	314	412	551	298
	separation at	1431	1332	1958	1164	1761

The significance of counterions present in the solvent for maintaining the double-stranded structure of DNA was seen in the free simulations already. Following this line, the stretching of microhydrated double-stranded DNA was investigated with a smaller number of counterions than required to neutralize the negative charge of DNA backbone (see the above discussion of such an occurrence). Performed were two further simulations for each DNA oligomer studied:

1. with 1/10 of the bulk number of water molecules (Dry1) and 1/2 of the number of counterions required for neutralization, and

2. with 1/20 of original water (Dry2) and 3/4 of the original counterions.

The stretching profiles for A_{13} are presented in figure 4.13. The influence of the decreased number of counterions is visible here. The overstretching transition occurs at a smaller external force than in the corresponding simulation with the same amount of water and the original number of counterions (passing from the red curve to the blue, and from the green one to the yellow). This effect was observed for all DNA species studied.

The observations in stretching simulations are summarized in table 4.7. Certain trends are visible even though only one simulation was performed for each DNA species and considered conditions. So, larger force is needed to induce the over-stretching transition as well as the separation of strands in the microhydrated systems Dry1 and Dry2, compared to the fully hydrated DNA. The effect of the content of counterions is more blurred, yet still the onset of the overstretched ladder structure seems to occur at a smaller force if some counterions are removed from the microhydrated DNA complex.

4.7   Conclusion

In this chapter, the structural features of microhydrated DNA and their changes upon stretching stress were investigated with classical MD simulation. The preliminary nature of this study does not allow to draw any quantitative conclusions. Still, there are several points to be made.

Both water and counterions are structure-stabilizing factors. The amount of water and the content of counterions necessary to support a helical double-stranded structure in the simulations was characterized. The removal of a certain number of counterions may stabilize the structure of a microhydrated DNA oligomer, even though such a system bears a non-zero electric charge.

Conformational transitions were observed at smaller forces when the DNA was stretched more slowly. This illustrates the irreversibility of the stretching process in the simulations, when the pulling is several orders of magnitude faster than in experiments. For this reason, further interpretation of the stretching simulations is limited to the comparison of the response of microhydrated DNA with the response of DNA in solution, pulled with the same loading rates.

The helical structure of DNA undergoes marked changes upon removal of water from the cluster. First, the helix bends toward the major groove where counterions accumulate, or passes to a barrel-like structure. Then, the double strand loses its helical character roughly at 6 water molecules per phosphate group. Continued drying induces a rupture of hydrogen bonds and a collapse of the double-stranded structure to a disordered compact coil.

The helical structure of microhydrated DNA can be supported by pulling at the 3′-ends. When microhydrated DNA is pulled by an external force, conformational changes require larger force than those in fully hydrated DNA. This may be explained by the tendency of a molecular cluster in vacuo to maintain a compact shape, against which the pulling force acts.

When a part of the counterions is removed from the microhydrated system (provided this situation is real), the overstretched ladder structure occurs at a weaker external pulling force.

In conclusion, extensive investigation of the structural changes of DNA under conditions resembling the experiment was shown for homogeneous DNA sequences. The developed simulation strategies for the stretching of DNA and the simulation of microhydrated DNA constitute the basis for the investigation of charge transport and charge transfer under these conditions in the next chapters.

CHAPTER 5

Charge Transport in Stretched DNA

In this chapter the mechanically stretched DNA sequences will be tested for their charge transport capabilities. Stretching is a common feature of the single-molecule experiments where the current is measured as a function of electrode distance. Simulations of the CT in those stretched DNA molecules are able to provide deeper insight into the processes involved in this experimental setup. Interpretation of the outcome of these experiments is possible based on the shown results. Moreover, choices of the DNA sequence for nano-electronic applications can be given.

5.1 Investigated Sequences and Structures

To gain deeper insight into the effect of stretching on the CT, seven different DNA oligomers are investigated in detail (complementary strand not shown): Three oligomers with purine nucleobases located exclusively on one strand in the cen-

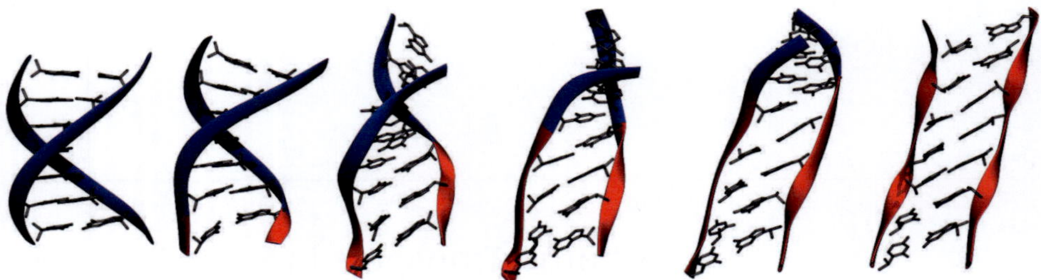

Figure 5.1: *The central seven base pairs of the DNA oligomer GG being stretched up to the additional extension of ca. 25 % of the free length by pulling at the 3'-ends. The backbones are depicted as ribbons, with color coding the conformation: blue – helical, B-like dsDNA; red – ladder-like S-DNA. The helical dsDNA structure passes to a ladder-like one, gradually.*

tral fragment, 5'-(G)$_{13}$-3' (GG), 5'-GG(AG)$_4$G-3' (GA) and 5'-CC(A)$_9$CC-3' (AA). And four oligomers with purines distributed among both strands, 5'-GG(TG)$_5$G-3' (GT), 5'-GG(CG)$_4$G-3' (GC), 5'-GG(AT)$_3$AGG-3' (AT) and 5'-CGCGAATTCGCG-3' (DD, Dickerson's dodecamer adopted from PDB ID 1BNA).[108]

Stretching simulations, as described in chapter 4, were performed for these seven oligomers. From the stretching simulations with linear increasing force, sequences of 14 to 19 (depending on DNA species) snapshots of the structure were taken. The snapshots were taken starting with the canonical helical structure until just after the conformational change to the ladder-like structure. Representative snapshots of the corresponding structures are shown in figure 5.1.

The length of the central fragment of seven base pairs (eight in case of DD) of theses sequences started at ca. 90 % of the equilibrium length observed in the free NPT simulation, see Tab. 5.1, and increased in steps of 0.1 nm.

Starting from these snapshot structures with different lengths MD simulation of 20 ns were performed. In these simulations the length of the central fragment was restrained by means of a harmonic potential acting between the 3'-terminal O3'-atoms of the central fragment with a force constant of 1000 kcal $\cdot$ mol$^{-1}\cdot$ nm^{-2}, or 6,948 pN/nm. The first 5 ns of these simulations were discarded, and the remaining 15 ns were taken for further analysis.

In the stretching simulations, all of the DNA species go through a transition to a ladder-like conformation (S-DNA).[21] The transition is gradual, meaning that one

Table 5.1: *Mean distance of the 3'-terminal O3' atoms in the studied DNA oligomers in free MD simulations (nm). Note that the distance is larger than B-DNA helix length (2.38 nm and 2.72 nm for the heptamers and the octamer DD, respectively) as the O3'–O3' vector is not parallel to the helical axis.*

sequence	length	sequence	length
GG	2.89	GT	2.76
AA	2.57	GC	2.69
GA	2.69	AT	2.78
		DD	3.17

or two base-pair steps at one terminus of the double strand pass to an S-like conformation first, followed by further neighboring base-pair steps and so forth, so that a mixed B-like/S-like structure appears. Eventually, the entire double strand ends up in the S-DNA structure. This process is shown in Fig. 5.1. The DNA double helix is unwound gradually, while the interstrand hydrogen bonding remains preserved between the complementary nucleobases during the entire stretching process. No deeper structural changes are observed, like e.g. swiveling of nucleobases out of the stack.

5.2 Charge Transport Calculations

The electric current was evaluated using the Landauer–Büttiker framework (see chapter 2.5.1) as a function of the varying end-to-end distance, for each of the studied DNA species. At every elongation, 7,500 snapshots were picked from the last 15 ns of the simulations with constant length. On each of these snapshots an individual calculation of the current was performed. The obtained distributions of current are highly asymmetric, as it turns out that the major portion of the current is contributed by a small number of conformations. This phenomenon was observed in previous works.[70, 81, 123]

The resulting mean values of the current are presented in Fig. 5.2. Apparently, CT in the two kinds of DNA sequences (those with all purines on one strand and those with purines distributed among both strands) responds differently to mod-

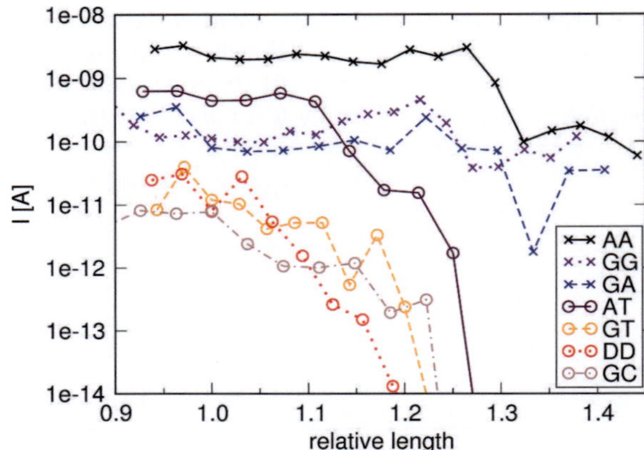

Figure 5.2: *Hole transport efficiency expressed as electric current at high voltage through the studied DNA devices depending on the end-to-end distance of the DNA strand. The distance is given relative to the equilibrium length of the respective DNA oligomer in a free MD simulation.*

erate stretching. There is little change of current through DNA species with all purine nucleobases on one strand (GG, AA and GA) until the elongation of ca. 30 %. The other DNA oligomers, with 'mixed' nucleobase sequences, show a steep decrease of conductivity, starting at the elongation of 10 to 20 %. This can be compared with the dependence of the measured current passing through DNA on the distance of electrodes, reported in several experiments.[8–10, 12] In all of these, the stretching of the DNA species led to an attenuation of conductivity at some point, which occurred within a narrow interval of the end-to-end distance. The authors explained these observations in terms of the DNA species detaching from the electrodes (possibly consecutively, if several DNA molecules had been bound between the electrodes and were detaching in a sequential manner).

An alternative interpretation can be proposed on the basis of the simulation results: The point at which the conductivity of DNA of mixed sequence drops steeply may correspond to the onset of overstretching transition, rather than a detachment of DNA from an electrode or another crude structural defect. Importantly, this 'breaking point' does not require an overstretching transition to be completed, and occurs at much smaller relative elongations – of ca. 20 % for the quite short DNA oligomers studied here, and arguably even smaller for longer oligomers.

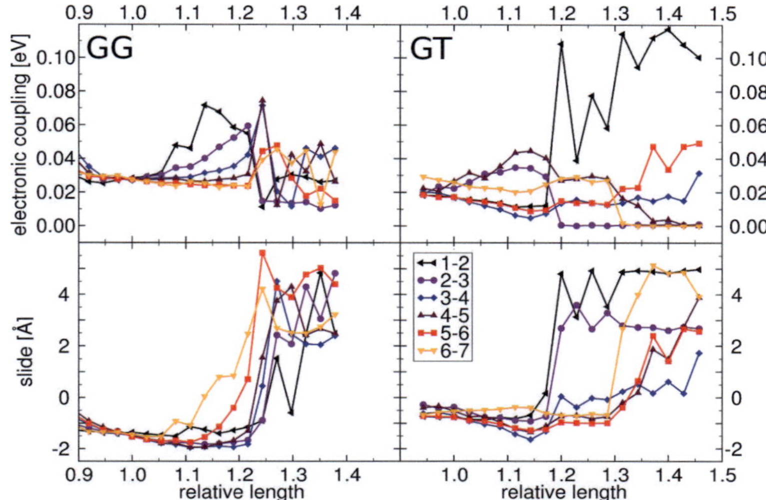

Figure 5.3: *Electronic coupling (EC) and value of the helical parameter slide for the individual base-pair steps, for the sequences GG (top) and GT (center). Mean values for ensembles generated by MD simulations are presented; see Tab. 5.2 for the designation of base-pair steps.*

5.3 Sequence Dependent Charge Transport

The aim here is to find out why the CT in different DNA species responds to stretching differently. To do so, the electronic couplings for the individual base-pair steps are analyzed. See the comparison of DNA species GG and GT in Fig. 5.3. Evidently, the electronic coupling for the base-pair step 2–3 vanishes in GT at the elongation of 20 % and does not recover any more. This happens in accordance with the current dropping to the sub-picoampere range at the same elongation.

Such a notable change of the electronic coupling has to originate in an abrupt conformational transition. Indeed, the elongation of 20 % is the point at which the transition to a ladder-like DNA structure starts. Note that while the conformational transition involving one or two base-pair steps occurs at the elongation of 20 % for the short dsDNA oligomers studied here. It may correspond to smaller relative elongation if longer DNA species are considered. The base-pair step 2–3 is one of the first two to take part in it, as illustrated by the sudden increase of the helical parameter slide. Fig. 5.4 gives a short reminder of the meaning of the helical parameter slide.

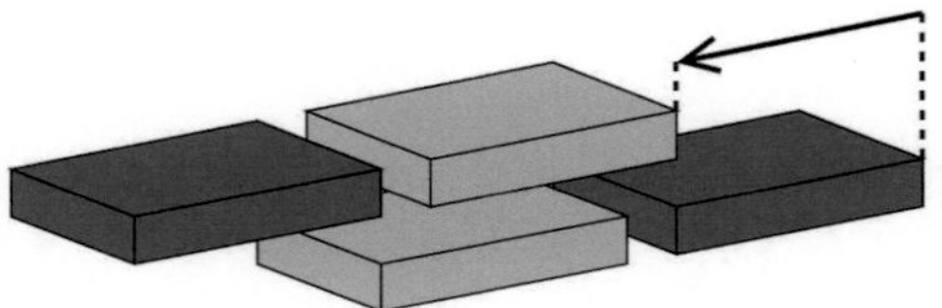

Figure 5.4: *Meaning of the helical parameter slide; every box depicts one nucleobase.*

However, this justification does not hold for every base-pair step. For instance, the base-pair step 1–2 in GT undergoes the same conformational change, and there is even a large increase of the corresponding electronic coupling. The analysis of a sequence with all purine nucleobases on one strand (GG) reveals a similar conformational transition, with an increase of slide for several base-pair steps. The corresponding electronic couplings decrease though, but retain non-zero values unlike those in GT. Consequently, the conductivity of GG decreases only slightly. This accumulated evidence prevents simple conclusions to be drawn, and a more detailed explanation has to be sought.

5.4 Detailed Structural Differences

To understand the different response of CT in the various DNA sequences to stretching, it is necessary to take into account both the inherent asymmetry of the DNA double helix and the particular way that the ends of the strands are pulled. In agreement with the mentioned experiments,[8–12] this study considers anchoring of the 3′-ends of the DNA double strand, see the illustration in Fig. 5.5.

When a DNA species with all purine nucleobases on one strand is stretched up to a ladder-like structure (Fig. 5.5 left), the overlap of every two neighboring purines decreases, as described by slide taking large values. The electronic coupling of purines in such an arrangement decreases though, but does not vanish. The situation is very different when the purine nucleobases alternate between both strands (Fig. 5.5 right).

Table 5.2 shows an overview of the base-pair steps present in the investigated sequences. Basically three different step types can be distinguished. For example, G | G is the intrastrand base-pair step of two GC base-pairs where the guanines are on top of each other bound to the same strand. G\T is a base-pair step where

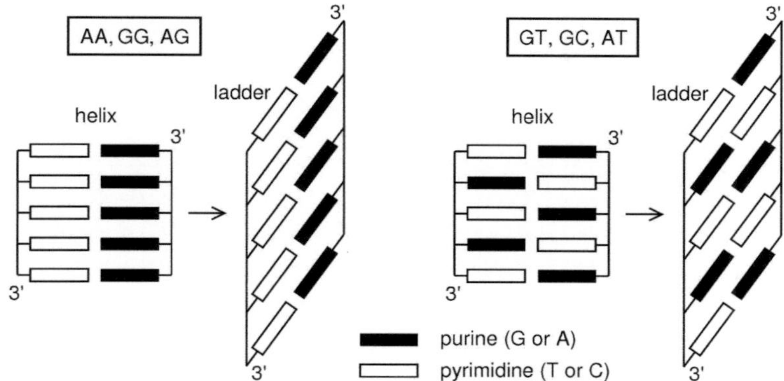

Figure 5.5: *Schematic depiction of the change of relative orientation of purine nucleobases (G, A) upon the onset of overstretched, ladder-like structure in a dsDNA fragment.*

a GC and a AT pair lay on top of each other where the guanine is bound to on strand and the adenine is bound to the other one. This situation corresponds to a purine–pyrimidine base-pair step. In G/T the situation is the same, but directionally inverted. So, this corresponds to a pyrimidine–purine base-pair step. For a graphical illustration of this naming scheme see figure 5.6.

In a purine–pyrimidine base-pair step, the purines actually approach to a very short distance in a ladder-like DNA structure, which can bring on a largely increased electronic coupling. This is what is observed for the 1–2 base-pair step in the GT sequence.

However, the effect is the opposite for the pyrimidine–purine base-pair steps. Here, the lateral distance between the purines increases markedly, making their overlap attenuate to a negligible value, so that the electronic coupling vanishes effectively. A previous computational study based on static DNA structures proposed a possible orientation of molecular orbitals in overstretched DNA in this way.[38]

This observation is the same for the other studied DNA species with purine nucleobases distributed among both DNA strands, too. In each case, there is at least one pyrimidine–purine base-pair step, which undergoes the described conformational change accompanied with vanishing electronic coupling, leading to a dramatic decrease of CT efficiency.

Table 5.2: Base-pair steps in the considered DNA sequences. $X\backslash Y$ – purine–pyrimidine step, X/Y – pyrimidine–purine step, $X\,|\,Y$ – purine–purine step.

			DNA sequence				
step	GG	AA	GA	GT	GC	AT	DD
1–2	G|G	A|A	G|A	G\A	G\G	A\A	G/G
2–3	G|G	A|A	A|G	A/G	G/G	A/A	G|A
3–4	G|G	A|A	G|A	G\A	G\G	A\A	A|A
4–5	G|G	A|A	A|G	A/G	G/G	A/A	A\A
5–6	G|G	A|A	G|A	G\A	G\G	A\A	A|A
6–7	G|G	A|A	A|G	A/G	G/G	A/A	A|G
7–8							G/G

Figure 5.6: Graphical illustration of the different base-pair steps in the DNA sequences.

5.5 Conclusion

In this chapter, the charge transport in the Landauer–Büttiker picture in stretched DNA was investigated. This setup resembles common charge transport experiments, where the DNA strand is attached to electrodes with varying distances.

The efficiency of hole transport in the studied dsDNA oligomers is little affected by stretching of the molecule by up to 10 % of the free length. When the DNA molecule is stretched further, up to the elongation of 30 %, the response of CT efficiency is determined by the exact nucleobase sequence. This effect has been characterized for the pulling of DNA at 3'-ends, in detail. The induced conformational changes lead to a breaking point, where the distance between purine nucleobases increases within pyrimidine–purine base-pair steps largely. The corresponding electronic coupling vanishes and the overall CT efficiency drops steeply. The first conformational transition of a base-pair step has been found the limiting parameter. While this conformational transition occurs at an elongation of ca. 20 % here, it may occur at smaller relative elongations in longer DNA species.

These results provide a means for the interpretation of the outcome of conductivity experiments where the DNA molecule is stretched, provided the DNA sequence contains pyrimidine–purine steps, which is actually the case in the mentioned experiments. The accumulated knowledge makes it possible to construct DNA sequences such that the extent to which their conductivity is affected by the mechanical stress during the experiments is controlled. The suggestion for DNA species attached with its 3′-ends to the electrodes would be to avoid pyrimidine–purine steps in the nucleobase sequence. Suitable DNA sequences can be opted for in the design of DNA-based devices as nano-scale electronic elements, where limited dependence of conductivity on length is desirable.

CHAPTER **6**

Charge Transport in Microhydrated DNA

Reproduced in part with permission from
M. Wolter, M. Elstner and T. Kubař,
J. Chem. Phys., 139, 125102 (2013).
Copyright 2013, American Institute of Physics.

In this chapter, the CT properties of DNA in environments with decreased amount of water will be investigated. Generally, the CT experiments are not conducted under "dry" conditions. The DNA strands are attached to electrodes and then dried in some fashion to prevent measurements of the conductivity of the solvent. In this setup, there is no way to tell how many water molecules stay attached to the DNA strand. Surely, the inner water molecules, tightly bound by hydrogen bonds, will not be blown away by a stream of nitrogen. Therefore, the described simulation setup will give insight into the effect this microhydrated state has on the the CT properties of DNA.

6.1 Investigated Sequences and Structures

Again, seven different DNA oligomers will be investigated in detail (complementary strand not shown): Three oligomers with purine nucleobases located exclusively on one strand in the central fragment, $5'\text{-}(G)_{13}\text{-}3'$ (GG), $5'\text{-}GG(AG)_4G\text{-}3'$ (GA) and $5'\text{-}CC(A)_9CC\text{-}3'$ (AA). And four oligomers with purines distributed among both strands, $5'\text{-}GG(TG)_5G\text{-}3'$ (GT), $5'\text{-}GG(CG)_4G\text{-}3'$ (GC), $5'\text{-}GG(AT)_3AGG\text{-}3'$ (AT) and $5'\text{-}CGCGAATTCGCG\text{-}3'$ (DD, Dickerson's dodecamer adopted from PDB ID 1BNA).[108]

As shown in chapter 4, the DNA structure deforms with the decreasing amount of water in the environment. For the Dry1 and Dry2 systems, the amount of water seems to be sufficient to preserve a regular double-stranded structure in all DNA sequences, while an applied harmonic restraint prevents the DNA oligo from bending towards the major groove. However, the helical structure of all of the DNA oligos is lost completely on hydration levels Dry3 and Dry4. Rather, a disordered compact coil appears, which cannot be remedied by the application of a simple restraint.

Nevertheless, all four kinds of microhydration environments will be analyzed in the following. Note, that only the data for Dry1 and Dry2 will relate to retained helical-like structures. This indicates that DNA does not form stable helices below a certain degree of solvation. Since the structure of DNA is vital for its conductivity, this is an important factor for the understanding of DNA conductivity experiments.

6.2   Charge Transfer Parameters

First, the averaged EC were obtained for all of the sequences and hydration environments along the MD trajectories. Figure 6.1 shows two examples of these mean values of EC for the different amounts of water. As a general trend the mean values of the EC are increased by 10 % to 150 % for the systems Dry1 relative to the respective fully hydrated systems. In most of the sequences, the EC decrease again in the systems with even smaller water content.

This is seen most markedly in the sequences AA, AT and AG. See figure 6.1 for sequence AT and the figures in the appendix B.2 for an overview of all investigated sequences.

In the sequence GT, the dependence of EC on the hydration level is different and distinct, as seen in Fig. 6.1. The difference between the base-pair steps GC-AT

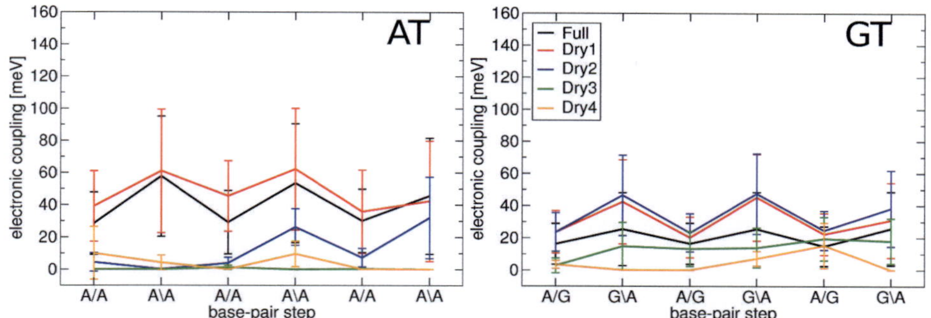

Figure 6.1: *Electronic coupling in the oligos At and GT, for fully hydrated system as well as the various levels of microhydration.*

Table 6.1: *Helical parameter rise (distance of stacked base pairs, Å units) in MD simulations of fully solvated and microsolvated DNA oligonucleotides.*

Oligo	AA	AG	AT	GG	GC	GT
Full	3.36	3.38	3.29	3.39	3.35	3.34
Dry1	3.28	3.34	3.23	3.36	3.34	3.27
Dry2	2.93	3.42	2.81	3.39	3.43	3.34

and AT-GC, discussed in a previous work,[124] is even emphasized in the low-humidity environment. Note, that the absolute values of the EC for fully hydrated systems in this work are ca. 30% smaller than in the cited previous work due to the orthogonalization of the basis set.

The EC are linked to the structure of DNA, and are particularly sensitive to the distance between nucleobases, which is roughly described by the helical parameter rise. Indeed, the rise decreases slightly for all of the Dry1 systems, and increases again for Dry2 in most of the cases, see Tab. 6.1. This may explain the observed differences of the EC between the various hydration levels. Clearly, the structural change due to dehydration impacts the electronic properties of the molecules. The behavior of the other parameters upon drying is inconclusive. See appendix B.1 for an complete overview of some representative helical parameters for all sequences under microhydrated conditions.

The energetics of hole transport in DNA is determined by the IP of the nucleobases. Apart from the mean value of the IP, the hole transport is determined by the fluctuations of the IP in time.

The std. deviation of the IP was found to be 0.3–0.4 eV in previous studies of fully solvated DNA.[124, 125] An important observation for the microhydrated systems is that the magnitude of these IP fluctuations is identical to that in fully solvated DNA. The std. deviation of the IP is 0.29–0.30 eV for the systems Dry1 and Dry2, as compared to 0.31 eV obtained for the fully hydrated system in this work. See appendix B.3 for an complete overview of the IP of the single nucleobases in the investigated sequences. Clearly, the fluctuations of the IP in fully hydrated DNA are caused by the nearest hydration shell(s) of the nucleobases – the water molecules in contact or very close to the nucleobases.

Table 6.2: ESP (in V) induced at a nucleobase in the GT oligo by the various parts of the molecular environment.

Environment	ESP induced by		
	DNA	water	Na^+
Full	−25.25	+6.91	+18.85
Dry1	−26.91	+0.76	+26.25
Dry2	−28.19	+0.84	+27.45

The IP of nucleobases are driven by the electric field induced by the molecular environment [124, 125]. Therefore, the electrostatic potential (ESP) induced by the environment on a nucleobase in the variously hydrated GT oligo was calculated.[124] Much the same as the fluctuations of the IP, the fluctuations of ESP remain the same even upon dehydration. Amounting to 0.30, 0.29 and 0.29 V for the fully hydrated, Dry1 and Dry2 systems, respectively. This is an evidence that the decisive portion of electric effect comes from the nearest hydration shell of the nucleobases.

Further, the ESP on a nucleobase was decomposed to contributions induced by the components of the molecular environment, which are DNA backbones with all of the other nucleobases, water molecules, and Na^+ ions. See Tab. 6.2 for an overview of the ESP of these groups. Interestingly, the magnitude of ESP induced by the ions increases, as does the ESP induced by the DNA backbones (so that it is more negative).

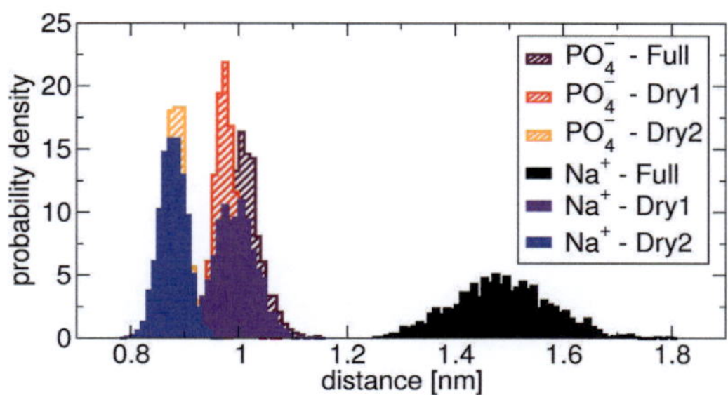

Figure 6.2: *Distributions of the distances of Na$^+$ ions and phosphate groups from the nucleobases in the AA oligo.*

To explain this, the distances of the Na$^+$ ions as well as of the phosphate groups in the backbones (which are the only charged groups in the system) from the DNA helical axis were calculated. The helical axis was obtained for every MD snapshot with the SCHNAaP algorithm implemented in 3DNA.[107, 126] The probability distributions of these distances are shown in Fig. 6.2. See appendix B.5 for values of all groups as mean values over time and atoms for the different sequences. Apparently, as the amount of water in the system decreases, both the ions and the phosphate groups approach closer to the probed nucleobase. This is not unexpected for the Na$^+$ ions as these are confined to the DNA hydration shell, which becomes more restricted spatially upon the partial removal of water. Remarkably, the structure of the DNA changes in a way that the phosphate groups approach closer to the nucleobases, too, contributing to the overall ESP in the opposite way. As for the ESP component due to water, which reduces effectively upon dehydration, the most likely interpretation is that the DNA hydration shell is affected by the approaching Na$^+$ ions. Thus, the water molecules have a larger positive charge to shield, and the decreasing ESP corresponds actually to increasing negative polarization rather than depolarization.

Summarizing the findings so far, there are two distinct effects of dehydration on the electronic structure. First, the change of geometrical structure of DNA due to desolvation leads to a change of the EC, which are not affected by the solvent substantially. On the other hand, the IP are affected largely not by the structure of nucleobases and base pairs, rather mostly by the fluctuations of the ESP due to solvent dynamics. These solvent fluctuations were shown to have a major impact on DNA conductivity,[44] therefore, a similar solvent effect on Dry1 and Dry2 is expected as found for the fully hydrated DNA.[44]

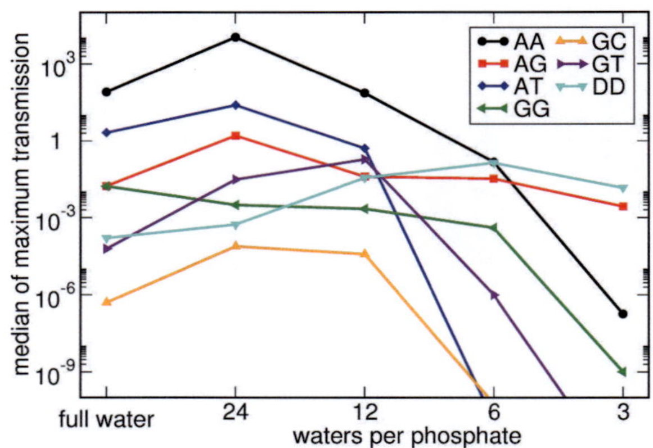

Figure 6.3: *Maximum of transmission function calculated with the Landauer–Büttiker method for all of the DNA oligos in the different microhydrated environments. (Median values obtained along 15 ns MD simulations.)*

6.3 Charge Transport Calculations

The model Hamiltonian containing the obtained CT parameters can be used as a basis for the approximative calculation of the electronic transmission function supported by the system within the Landauer–Büttiker framework.[127]

Here again, the previously developed multi-scale framework,[45] already successfully applied to fully hydrated DNA species,[44] was used.

The transmission function was evaluated for 7,500 snapshots along every MD simulation of 15 ns. The maximum value of the function was recorded for every snapshot, and the median value was taken to represent each distributions of max. transmission obtained in that way. Fig. 6.3 shows these median values obtained for each of the DNA sequences, in full aqueous solution as well as in the microhydrated systems Dry1 to Dry4.

The transmission increases in the Dry1 system in comparison to the fully hydrated one for five out of seven DNA oligos. This effect is less obvious for the systems Dry2, possibly due to the slightly distorted helical structure at this level of hydration. With further reduced water content in Dry3 and Dry4, DNA suffers from severe distortions and the transmission vanishes in five of the studied DNA species.

The increase of transmission in Dry1 systems can be explained in terms of the CT parameters. The EC increase for all Dry1 and some of the Dry2 systems, while other characteristics like the fluctuations of the IP remain the same. Starting at the microhydration level Dry2, the structural disorder makes the calculations of transmission impracticable.

The Landauer–Büttiker calculations as performed in this work were used to obtain a simple indicator of the molecular conductivity. Basically, the calculated transmissions represent a measure of structural and dynamical disorder of the linear conducting chain, assuming coherent transport. This assumes that CT occurs in a single-step fashion, i.e. a hopping picture of transport is excluded. The fact that this approximation does not capture the CT in DNA fully and that the time order of conformations (the true time dependence of CT parameters) does play a role, has been shown recently by using the time series of CT parameters as described above.[128] Still, even this improved description neglects the polarization of environment, its response to CT and the effect on CT energetics. Therefore, the Landauer transmissions described above as well as the improved time-dependent treatment in Ref. 128 are both merely indicative. A hopping picture would also have to involve on a more systematic footing the effect of environmental polarization, which can be quantified by means of (outer-sphere) reorganization energy (RE), one of the crucial parameters in Marcus' theory of electron transfer (see also chapter 2.6.1).[71, 85] RE is the energy required to transform the structure of the donor and acceptor molecules as well as of the environment from that corresponding to the initial charge state α to that of the final state β (after the transfer has taken place). The RE of the solvent represents the major contribution to the activation energy for CT and is a decisive factor determining the efficiency of CT in systems immersed in a polarizable medium; aqueous solution is a typical example. Detailed information on how to calculate the RE of DNA with MD simulations is given in the last part of chapter 2.6.1

The values of RE for the fully hydrated system are in accordance with previous works. With the described approach the λ_s for CT in DNA were found to be 1.21 eV for a adenine–adenine CT event and 1.41 eV for a guanine–guanine CT event. [93]

Interestingly, nearly identical values were obtained for the Dry1 system, with the difference smaller than 0.1 eV, see Fig. 6.4. In Dry2, the spread of the obtained values is huge due to structural distortions, and so the larger RE observed for

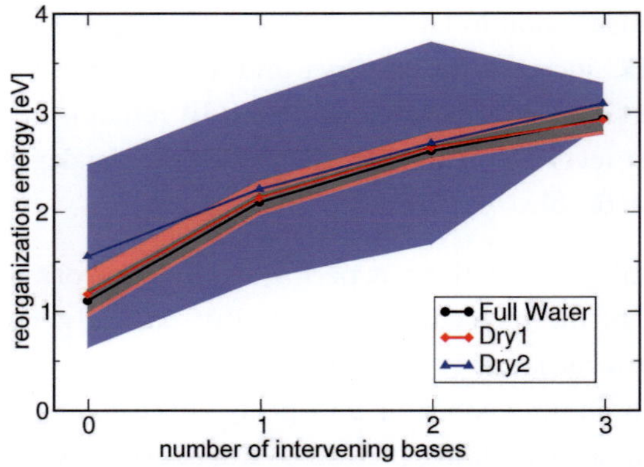

Figure 6.4: *Reorganization energy for hole transfer from one nucleobase to another across 0–3 intervening ones in the AA oligo in the various environments. (Shaded area – std. dev.)*

next neighbor transfer may be an undersampling artifact. It is truly astonishing to see that the RE is little affected by the removal of certain (large) amounts of water around the DNA. Still, the removal of the innermost hydration shell would presumably have a large impact on the RE. However, this is a setup for which it seems to be nearly impossible to achieve sufficient sampling in an MD simulation. Therefore, dried DNA (as long as not too much water is removed so that a stable helical structure is retained) seems to have very similar charge transport properties as fully solvated DNA. The EC and the IP are nearly identical, and the RE is very similar.

Therefore, a very similar conductivity (charge mobility) of DNA for either transport mechanism, tunneling or hopping, is expected.

6.4   Direct Dynamics of Charge Transfer

To prove this point, charge transfer simulations using a time-dependent scheme, were performed, thereby extending Landauer's coherent transport picture. See chapter 2.6.2 for details on the used charge transfer methods.

Here, hole transfer in the central pentanucleotide of the AG oligo was studied. Ten simulations of 1 ns were performed for the fully hydrated and the Dry1 systems

Table 6.3: Number of CT events per ns observed in non-adiabatic simulations of hole transfer in the AG oligo.

system	mean-field	surface hopping
Full	33 ± 5	45 ± 11
Dry1	20 ± 8	24 ± 8

twice, using different non-adiabatic propagators – the mean-field method and the surface hopping. The events of hole hopping from one G to another were counted and averaged over each group of ten simulations; the results are shown in table 6.3.

Very similar rates of CT are obtained with the two methods used. Although, the rate is decreased by ca. 45 % in the microhydrated system Dry1. In contrast, the electric transmission obtained with the Landauer model was larger in the Dry1 systems than in the fully hydrated ones, for most of the DNA oligos including AG. While this difference may surely be accounted to the entirely different character of the methods, another point should be emphasized here: The difference observed between the fully hydrated systems and the microhydrated Dry1 (and Dry2 where applicable) is actually smaller than what may have been expected, considering the very different structure of the systems – DNA in bulk solution vs. DNA in complex with water molecules, however in the gas phase.

6.5 Conclusion

In this chapter, the charge transport and charge transfer in microhydrated DNA strands was investigated. The idea of microhydrated DNA was induced by the experiments, where the DNA strand is dried before the measurements. However, the exact amount of water preserved at the DNA strand is unknown and cannot be determined by the experiments alone. The simulation under these conditions was able to resolve the question how these conditions affect the CT.

In conclusion, the efficiency of electron transfer / transport in microhydrated DNA that has retained regular double-stranded structure is similar to that in DNA in bulk solution. The magnitude as well as fluctuations of the electric field induced by the aqueous environment are due to the closest hydration shell(s) from the decisive part. The changes of the electric field upon dehydration were characterized and its components were shown to compensate to a large extent. The electronic couplings, ionization potentials as well as reorganization energies show rather moderate differences between the fully hydrated and the microhydrated systems. Therefore, it seems conceivable that the transport of electric charge in a microsolvated complex molecular system (e.g. molecular junctions) will exhibit similar characteristics as that in the same system in bulk solution, as long as the structure of the closest solvation shell is retained.

CHAPTER 7

A Parametrized Model to Simulate CT in DNA

In this chapter, a parametrized model for the charge transfer in DNA is presented. As shown in the previous chapters, the CT in DNA is sensitive to several quantities: the electronic couplings (EC), the ionization potentials (IP) and the influence of the environment. In the shown QM/MM setup, these parameters are obtained with a combination of classical MD simulations and QC calculations. Both of these can be very time consuming, particularly when it comes to large system sizes or extended periods of time.

The idea of the parametrized approach is to model these parameters without the need of MM or QC calculations. To do so, data acquired from QM/MM simulations are analyzed extensively in order to model the CT quantities with simple algorithms. As a first step, the EC and IP are parametrized and an appropriate Hamiltonian is modeled, which already can be used in charge transport calculations in the Landauer–Büttiker picture. The next step is to include the environmental effects in the model approach.

Here, the charge is introduced into the CT model, and the resulting polaron is modeled. Relaxation effects of the environment are taken into account in addition. Finally, first test calculations of the charge transfer in the DNA sequence GAG are performed.

7.1 Creating the Electronic Couplings

MD simulations were performed for DNA sequences containing all 10 possible base-pair steps. The EC were calculated in every MD step (1 fs) over a period of 100 ps, and the time series were analyzed to obtain the magnitudes and characteristic frequencies of the fluctuations.

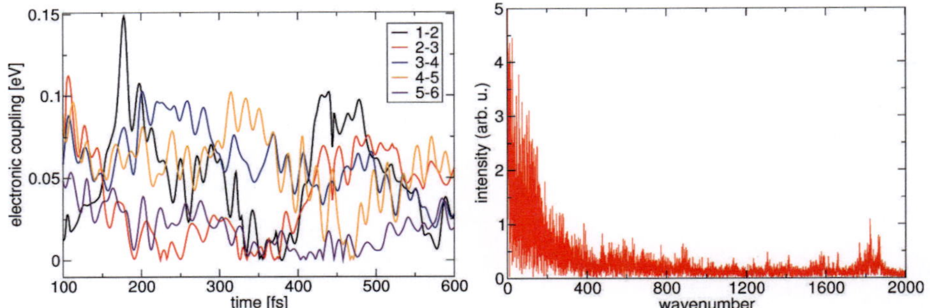

Figure 7.1: EC calculated from MD simulation of 100 ps. Left: Time series of EC shown for 500 fs calculated for 5 base-pair steps. Right: Power spectrum of the time series from 100 ps simulation, averaged over five base-pair steps.

Figure 7.1 shows the time series of the EC in an MD simulation of 500 fs calculated for five base-pair steps in a polyA sequence. These time series were analyzed in terms of a Fourier transformation over the whole 100 ps simulation time to obtain power spectra. To reduce the noise, all five power spectra were averaged to obtain the relevant peaks present in all base-pair steps.

At this point, no characteristic frequencies can be evaluated from this power spectrum. Therefore, a straightforward approach (e.g. to model the fluctuations as a sum of goniometric functions) does not seem to be feasible.

An alternative approach to parametrize the EC is to create random numbers with the correct statistics. To this end, EC time series were obtained from MD simulations of 20 ns with the EC calculated every 2 ps. Since the sign of the EC has no physical meaning, being dependent on the sign of the wave function, the obtained time series contain only absolute values of the EC. This leads to two probability distributions for the EC, one with a positive mean value and another one with a negative mean value. Hence, the resulting distribution of the absolute EC is a sum of two Gaussian-type functions.

So, each of the obtained probability distributions of EC of all ten possible base-pair steps was fitted with a function of the following form:

$$P(EC) = \frac{1}{\sigma\sqrt{2\pi}} \cdot \left(\exp\left[-\frac{(EC - \mu)^2}{2\sigma^2} \right] + \exp\left[-\frac{(-EC - \mu)^2}{2\sigma^2} \right] \right) \qquad (7.1)$$

Table 7.1 shows the resulting parameters σ, the width, and μ, the mean value, of the distributions.

Table 7.1: *Mean values and standard deviations of the EC estimated for the different base-pair steps.* $X\backslash Y$ – *purine–pyrimidine step,* X/Y – *pyrimidine–purine step,* $X\,|\,Y$ – *purine–purine step. See also figure 5.6 for a graphical illustration.*

base-pair step	A|A	A/A	A\A	A|G	G|A	G/A	A\G	G|G	G/G	G\G
σ [eV]	0.0388	0.0204	0.0473	0.0428	0.0375	0.0180	0.0287	0.0297	0.0086	0.0218
μ [eV]	0.0480	0.0480	0.0497	0.0022	0.0008	0.0059	0.0002	0.0182	0.0089	0.0148

Now, time series of EC with these probability distributions have to be created, to be used in the parametrized model. The approach presented here is to draw random values from these distributions. Therefore, the following algorithm was implemented into the CT model.

A standard linear congruential random number generator is applied, which generates series of uniformly distributed random number in the interval (0,1). These uniformly distributed random numbers are now transformed to a normal distribution.

The cumulative distribution function of the normal distribution has the form:

$$\Phi(x) = \frac{1}{\sqrt{2\pi}} \int_{-\infty}^{x} \exp\left[\frac{t^2}{2} \right] dt = \frac{1}{2}\left[1 + \operatorname{erf}\left(\frac{x}{\sqrt{2}} \right) \right] \qquad (7.2)$$

The inverse of this function, called the probit function, is used to get the distribution of the electronic couplings. The probit function can be expressed in terms of the inverse error function.

$$\Phi^{-1}(x) = \sqrt{2}\,\operatorname{erf}^{-1}(2x - 1) \qquad (7.3)$$

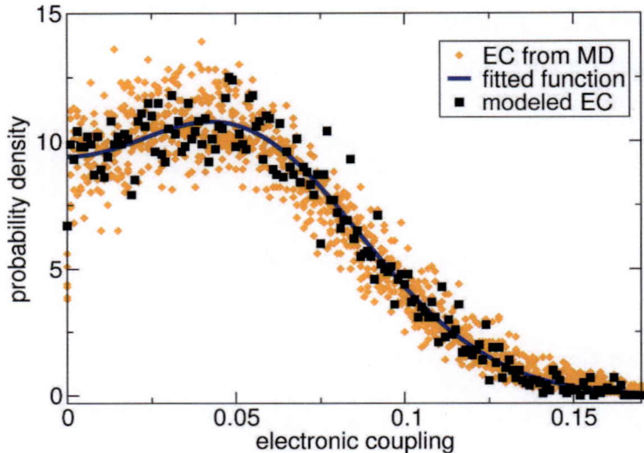

Figure 7.2: *Distribution of EC recorded during an MD simulation of 20 ns for six A | A base-pair steps (orange) and the fitted distribution function (blue). EC for one A | A base-pair step obtained from one CT model simulation of 20 ns are shown in comparison (black).*

where x is in the interval (0,1). Finally, the random value x is transformed with the obtained parameters σ and μ, giving a random value taken from the desired distribution:

$$F^{-1}(x) = \mu + \sigma \cdot \Phi^{-1}(x) \tag{7.4}$$

The calculation of the probit function is done with the algorithm by P. J. Acklam. [129] This algorithm has a relative error of less than $1.15 \cdot 10^{-9}$.

Figure 7.2 shows the distribution of recorded values of EC during an MD simulation of 20 ns for six adenine-adenine base-pair steps and the fitted distribution function (equation 7.1). EC for one A | A base-pair step obtained from one model CT simulation of 20 ns are shown in comparison.

Drawing random numbers in every step of the calculation has a fatal consequence: The calculated time series are not continuous. To prevent these unrealistic sudden changes, a new EC is generated in the described fashion only every 10 steps (10 fs). In between these steps, the EC is interpolated linearly. This might lead to distributions that deviate from the desired one slightly. Still, it is an acceptable first approach, and all test simulations are done with this possibly inaccurate setup.

7.2 Modeling the Ionization Potentials

MD simulations of 100 ps of polyA and polyG sequences were performed, and the CT parameters were calculated every 1 fs. The obtained time series of IP were analyzed in terms of a Fourier transformation to obtain a power spectrum. The spectra reveal basically two separate regions. The first are the wavenumbers above 1000 cm^{-1}, which correspond to the internal structural fluctuations of the nucleobases. Note that these fluctuations are slightly different for adenine and guanine. (See figure 7.3 for the comparison). Guanine reveals a few more characteristic frequencies in the region of 1800 cm^{-1}. The other region contains the characteristic frequencies due to the dynamics of the environment, i.e. the solvent and DNA backbone, which can be found below 1000 cm^{-1}.

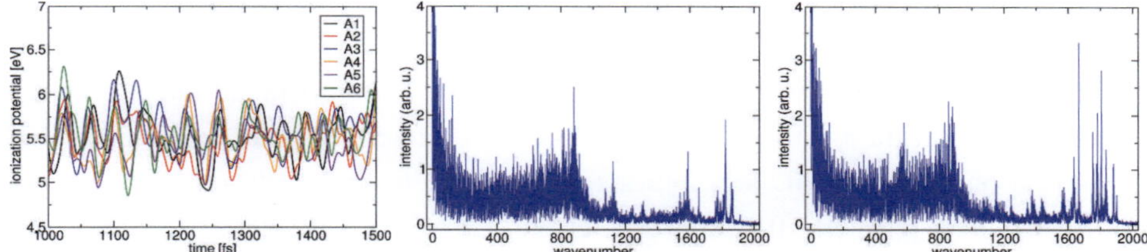

Figure 7.3: *IP calculated from MD simulation of 100 ps. Left: Time series of IP shown for 500 fs calculated for 6 adenines. Middle: Power spectrum of the time series of the IP of a adenine. Right: Power spectrum of the time series of the IP of a guanine.*

The most prominent peaks in both regions of these power spectra are identified and taken as characteristic frequencies. This way, 15 and 24 characteristic frequencies were found for the internal (>1000 cm^{-1}) and for the environmental fluctuations, respectively. Table 7.2 shows an overview of these obtained wavenumbers.

While the wavenumbers for internal fluctuations of the adenine and guanine nucleobases reveal certain differences, the wavenumbers obtained for the environmental fluctuations are the same for both nucleobases. The very low wavenumbers of the environment have been found crucial to obtain the correct probability distributions. In detail, the rare events where the IP is very high or very low can only be sampled with this wavenumbers included.

Having evaluated the characteristic frequencies, a sum of cosine functions was created to model the time course of the IP.

Table 7.2: Obtained wavenumbers for the fluctuation of IP due to internal dynamics of the nucleobases adenine and guanine and due to environmental dynamics

	Adenine	Guanine	Environment
obtained wavenumbers [cm^{-1}]	1866.7, 1860.0, 1823.3, 1790.0, 1770.0, 1756.7, 1653.3, 1583.3, 1540.0, 1380.0, 1310.0, 1246.7, 1126.7, 1076.7, 943.3,	1900.0, 1883.3, 1833.3, 1810.0, 1783.3, 1756.7, 1670.0, 1630.0, 1560.0, 1433.3, 1376.7, 1243.3, 1156.7, 1076.7, 1026.7,	883.3, 796.7, 733.3, 620.0, 450.0, 376.7, 323.3, 220.0, 176.7, 132.7, 103.3, 61.7, 41.7, 30.0, 20.7, 14.7, 9.7, 8.7, 4.7, 4.0, 3.0, 1.6, 0.7, 0.3,

$$IP(t) = \sum_n A_n \cdot \cos\left(\frac{t \cdot 2\pi}{T_n}\right) + E_0 \qquad (7.5)$$

T_n is the period and A_n is the corresponding amplitude of the oscillation.

While the T_n are obtained with the Fourier transformation, the A_n have to be obtained from other data. The amplitudes have to be fitted to reproduce the standard deviations of the IP in MD simulations accurately. To this end, additional simulations of 20 ns with the IP calculated every 2 ps were performed to obtain the statistics of the IP time series. Two types of simulations were performed to obtain different amplitudes for the internal and external fluctuations.

To obtain the amplitudes of the internal fluctuations, simulations with no QM/MM coupling were performed. This way, the IP will fluctuate only as a result of the internal modes of vibration of the nucleobases. Table 7.3 (left) shows the statistics of the IP of adenine nucleobases in a polyA sequence, when no QM/MM coupling is applied.

Now, the sum of cosine functions was set up including only the internal fluctuations with the corresponding periods. The sum of cosine functions was then shifted by a constant E_0 to yield the same mean values as obtained from MD simulations. And, the amplitudes A_n were chosen in a way that an overall magnitude of the fluctuations is achieved, which matches the value obtained from MD simulations without QM/MM coupling, i.e. 0.1 eV.

To obtain the amplitudes for the environmental fluctuations, simulations with QM/MM coupling were performed. Again, the obtained time series of IP were analyzed for mean values and standard deviations. The mean values showed that

Table 7.3: Statistics of the IP of adenine nucleobases in a polyA oligomer. Left: Simulation without QM/MM coupling; only internal modes of vibration contribute to the fluctuations. Right: Simulation with QM/MM coupling to the environment; fluctuations of solvent and DNA backbone contribute to the fluctuations of IP.

in vacuo		QM/MM	
nucleobase	mean $\pm$ std. dev. [eV]	nucleobase	mean $\pm$ std. dev. [eV]
A_1	5.354 ± 0.099	A_1	5.565 ± 0.304
A_2	5.356 ± 0.090	A_2	5.572 ± 0.303
A_3	5.357 ± 0.101	A_3	5.563 ± 0.301
A_4	5.357 ± 0.095	A_4	5.554 ± 0.302
A_5	5.357 ± 0.095	A_5	5.549 ± 0.304
A_6	5.359 ± 0.096	A_6	5.551 ± 0.305

an additional shift of IP had to be introduced to account for the environmental effect. Then the amplitudes A_n of the cosine functions of the environmental fluctuations were fitted in combination with the amplitudes for the internal fluctuations, obtained previously, so that the larger fluctuations of about 0.3 eV observed in the QM/MM environment are reproduced.

Additionally, the amplitudes of the cosine functions that correspond to the nine lowest environmental fluctuations were scaled down by a factor of 0.6. These slow fluctuations correspond to movements of the whole molecular system, therefore they have very long periods and rather low amplitudes.

The MD simulations performed to obtain the time series of IP were conducted using a non-polarizable force field. In such simulations, the electric field induced by the environment is overestimated by 26-34% [80], therefore QM/MM interactions have to be scaled down by a factor of 1.5 for charge transfer calculations[73]. To model this scaling, MD simulations with and without this scaling factor were performed and the CT parameters recorded. The time series were analyzed for mean values and standard deviations. A new set of amplitudes for the environmental fluctuations was introduced to reproduce these scaled fluctuations.

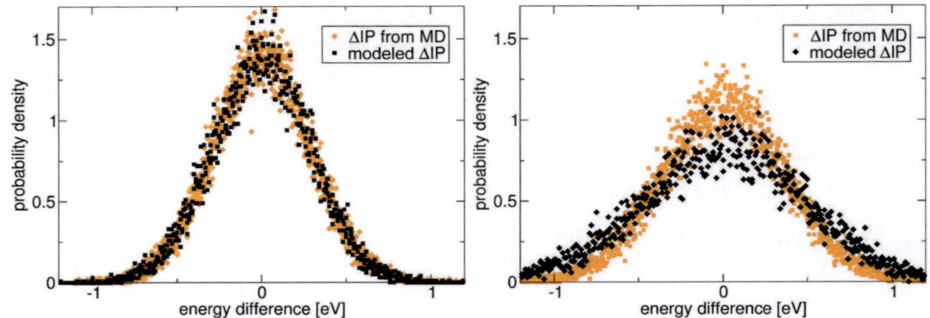

Figure 7.4: *Distributions of IP differences between neighboring bases in a polyA sequence. Shown are the distributions for the IP difference of nucleobases A_2-A_1 (left) and A_3-A_1 (right) obtained from an atomistic MD simulation (orange) and from the modeled simulation.*

Finally, it is important to make sure that the time series of IP is not the same in every simulation. Pilot tests with different random phase shifts for every internal and environmental fluctuation showed huge fluctuations of the resulting statistics. Due to randomly generated interferences of the cosine functions, both mean values and standard deviations took different values in every conducted simulation. An improved approach is to start the time series of IP at a randomly chosen starting point. For this, a single random number is drawn and set as a phase shift for the entire sum of IP generating cosine functions. This effectively solves the described issue occurring when using a whole set of random numbers.

Correlation of the Neighboring Nucleobases

The IP of neighboring nucleobases are correlated[124]. This is caused by the fact that all nucleobases are influenced by the same environment, and they are physically close to each other. MD simulations show that IP of neighboring bases have a correlation coefficient of 0.5 up to 0.7 while the correlation coefficient is still 0.3 for second neighbors. To model these correlations, a phase shift of the environmental fluctuations was introduced for the individual nucleobases.

The magnitude of the necessary phase shift can be obtained by comparison with the correlation coefficients obtained from MD simulations. Unfortunately, pilot calculations showed that this approach does not lead to reasonable phase shifts. A better parameter to monitor is the IP difference of the (neighboring) nucleobases.

Not only will this give a better understanding of the phase shift needed to obtain the correct correlations, but it is also a crucial parameter for the CT characteristics of the model. Therefore, IP differences between neighboring nucleobases were calculated from MD simulations of 20 ns and analyzed in terms of their probability distributions. The phase shift of the IP was parametrized so that the width of these distributions matches the data from MD simulations.

The final phase shift was set to 1.11 per neighbor for a cosine function with a period of 2π. Figure 7.4 shows the distributions for the first two neighboring nucleobases when this phase shift is applied. Note that the modeled standard deviation of the IP difference becomes inaccurate starting from the second-neighbor correlation. This will have an effect in approaches relying on the coherent transport, like the Landauer–Büttiker model. However, the CT model should work with charge transfer methods, which describe nearest neighbor hopping predominantly.

7.3 Testing with Charge Transport Calculations

With the EC and IP constructed, the time-dependent coarse-grained Hamiltonian can be set up, and the CT model can be tested with charge transport methods already. The easiest approach is to calculate the transmission function in every time step and average over the simulation.

As already noted, the CT model does not describe long-range correlations well. Thus, the calculation of a transmission function was done for a sequence of only three adenine nucleobases. Figure 7.5 shows the averaged transmission function in comparison to one obtained along an atomistic MD simulation. The overall form of the transmission function is reproduced accurately for this DNA species, apparently. For longer sequences, the transmission functions assumes a triangular shape as a consequence of a wrong distribution of IP differences starting from the A_1-A_4 neighbors. Clearly, this is an artifact of the simple correlation model.

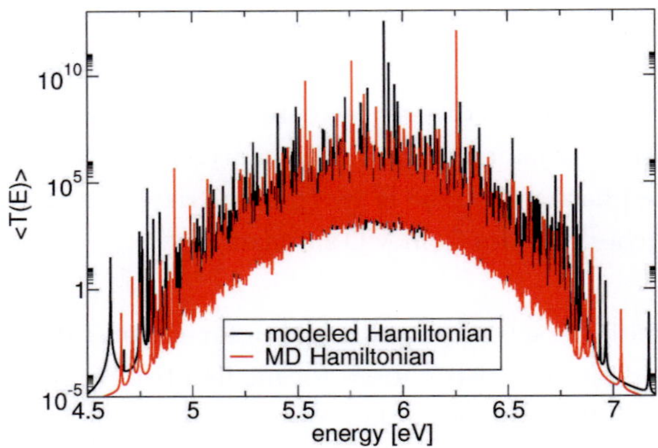

Figure 7.5: *Transmission function calculated with the Landauer–Büttiker model. Averaged transmission function for a MD simulation of 20 ns of a sequence of three adenines (red) and for the modeled CT parameters (black)*

7.4   Charge Transfer Extensions

In order to simulate charge transfer with non-adiabatic multi-scale schemes, certain additions to the CT model Hamiltonian described so far are necessary.

The Environmental Effect

Until now, the Hamiltonian was created for "neutral" systems, unaffected by a charge residing in the system. The introduction of a charge has several effects on the system and therefore on the CT Hamiltonian.

Previous work showed that the charge (hole) is stabilized by the polarization of the surrounding water molecules. This can be described as a polaron [96–99]. The stabilization can be seen in the lowered IP of the nucleobases carrying a charge. To quantify the decrease of IP of a charged nucleobase, simulations with a stationary charge (hole) introduced into the system were performed. A polyA sequence was equilibrated with a charge residing completely localized on one nucleobase. Then, an MD simulation of 1 ns was performed, and the Hamiltonian was calculated in every time step. During this simulation, the charge was restricted to reside on one nucleobase and did not transfer. The resulting IP time series were recorded for every nucleobase, and mean values were calculated. These mean values show that the polarization induced by the charge on a nucleobase also influences the IP of the neighboring nucleobases. With increasing distance, this influence is decreasing. Table 7.4 shows the obtained mean values of IP for the charged nucleobase and the five neighboring nucleobases.

Table 7.4: Lowering of IP due to environment polarization. Charge (hole) is localized on nucleobase 0; 1-5 indicating neighboring nucleobases in increasing distance. See also figure 7.6

nucleobase	0	1	2	3	4	5
Δ IP [eV]	-3.373	-2.326	-1.344	-0.828	-0.463	-0.159

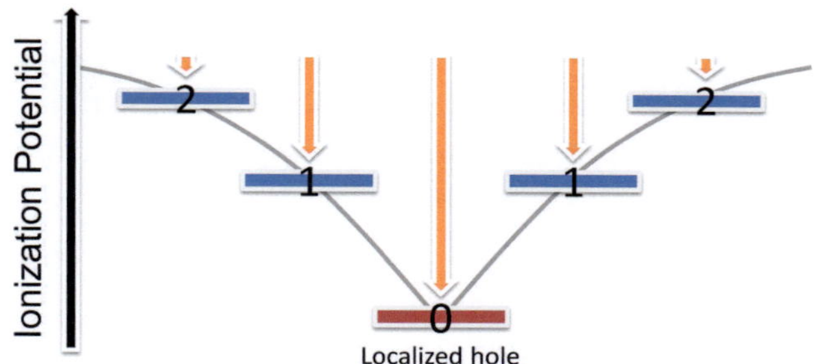

Figure 7.6: Graphical illustration of the decreased IP in a charged system. The charge is fully localized on the nucleobase in the middle (0, shown in red). Neighboring nucleobases (1 and 2, shown in blue) are influenced by the polarization induced by the charge on the nucleobase in the middle, albeit to a lesser extend

Now, to implement this effect, a shift of the modeled IP was introduced. In a first step, the IP are created according to the procedures shown above. Afterwards, these IP are lowered by the calculated energy shifts weighted by the charge populations on each nucleobase.

Relaxation Time

With the setup described so far, the environment re-polarizes instantaneously whenever a CT event has taken place. This behavior would neglect the fact that it takes a certain period of time for the environment to accommodate for the new charge distribution. Therefore, IP should not adjust instantaneously. To overcome this issue, a relaxation upon a CT event has to be implemented.

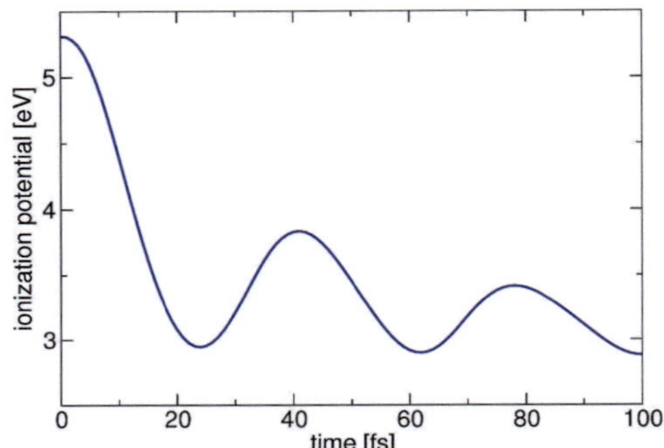

Figure 7.7: *Relaxation function of a fully localized charge hopping from one nucleobase to another. Function averaged over 1000 simulations of 100 fs each.*

To model this behavior, two alternative functions are available so far. First, a simple exponential function may model the change of IP of the nucleobases in a preset relaxation time. Second, to prevent a sudden change in IP at the onset of the exponential function, a Gaussian-shaped function may be implemented to represent the changes of polarization.

As an alternative to such relaxation functions, an empirical relaxation has been implemented. 1000 simulations of 100 fs length were performed with the IP calculated in every time step. PolyA sequences equilibrated with no charge on the nucleobases were simulated. Then, a charge was introduced into the system, and MD simulations were commenced. The time series of IP were recorded in all these simulations, and a mean value out of the 1000 simulations was calculated for every time step. Figure 7.7 shows the appearance of this averaged relaxation behavior. Averaging over 1000 simulations, the stochastic fluctuations of the system are canceled, and the resulting fluctuations of IP are an effect of the CT event. The obtained relaxation function has the form of a cosine function with an exponentially decreasing amplitude. There are maxima, located at roughly 40 and 80 fs. Until now, there is no arithmetical function for this behavior implemented, rather a table is read from a file. This has the advantage that any desired function can be used for the relaxation, increasing the flexibility of the approach.

The relaxation at a given time point is determined by an occupation history - the state of the charge is recorded for a chosen number of steps preceding the current step. Depending on this occupation history, the value of the relaxation function is determined and the IP is shifted accordingly.

The obtained relaxation function only covers a drop of the IP of about 2.5 eV within the initial 100 fs following a CT event. Comparison with the data in table 7.4 reveals a missing portion of the relaxation amounting to 0.9 eV, which has to occur for a fully localized charge. Until now, the obtained relaxation function is scaled up to account for the larger decrease of the IP. This clearly is an issue in the CT model that requires a solution. The work on the description of this additional relaxation on a longer time scale is underway presently.

The Hubbard Matrix

Also, a Hubbard matrix is needed for CT simulations. This matrix contains the relevant parameters to estimate the energy change due to different on-site electron–electrons interactions as well as due to relaxation and rotation of orbitals. In this CT model, the Hubbard matrix depends only on the input sequence. Therefore, it is set up only once at the beginning and is kept for the rest of the simulation. It is built from two components:

First, the diagonal elements are the Hubbard parameters of each nucleobase. Here, the Hubbard parameters of guanine and adenine are implemented as fixed values obtained from DFTB2 calculations; they amount to 5.66 eV for guanine and 5.58 eV for adenine.

Second, the off-diagonal elements take the form $\frac{1}{r_{ij}}$, where r_{ij} is the distance between the centers of mass (COM) of the neighboring nucleobases. These off-diagonal elements have been calculated as mean values for every possible nucleobase step. MD simulations of 20 ns were performed and the distances of the COM of the purine nucleobases were recorded. Table 7.5 shows the mean values for the ten different base-pair steps. Notably, difference is found between intra- and interstrand distances.

Table 7.5: *Mean values of the COM distances for the base-pair steps from 20 ns MD simulations. $X\backslash Y$ – purine–pyrimidine step, X/Y – pyrimidine–purine step, $X|Y$ – purine–purine step.*

Step	A|A	A/A	A\A	A|G	G|A	G/A	A\G	G|G	G/G	G\G
Distance [nm]	0.3851	0.6940	0.5221	0.4144	0.3783	0.5238	0.6364	0.4142	0.6724	0.5109

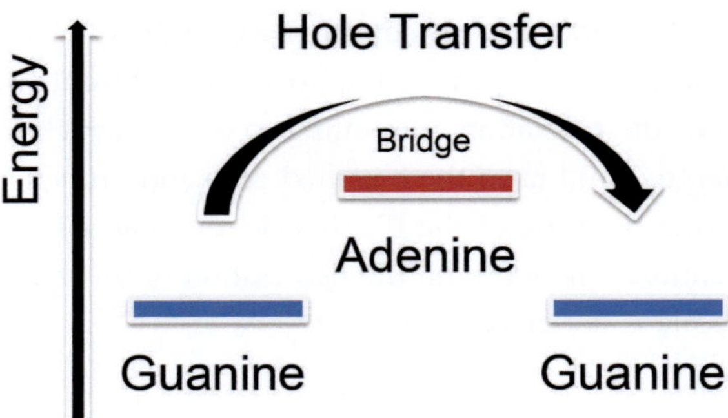

Figure 7.8: *Graphical illustration of the energy landscape in the GAG sequence. Guanines have the lowest IP. Adenine with higher IP acts as a barrier for the CT.*

Having set up the Hubbard matrix, it now has to be modified in terms of a self-interaction correction. All-DFT based methods suffer from the so-called self-interaction error. To correct this behavior, the second-order terms have to be scaled down. Therefore, the elements of the Hubbard matrix are scaled down with a factor of 0.2, the same as in the previous work [79].

Inner-Sphere Reorganization Energy

So far, the inner-sphere reorganization energy is missing in the CT model. Although this is not connected to the Hubbard matrix at all, it easily can be introduced by including it into it. Therefore, the inner-sphere RE, which was estimated to be 0.23 eV[79], is subtracted from the diagonal elements of the Hubbard matrix.

7.5 Testing with Charge Transfer Methods

At this point, the model is complete and should be able to describe charge transfer in DNA. To test this, hole transfer in the short DNA sequence GAG was simulated 100 times for 1 ns. This sequence was chosen because of its special characteristics for charge transfer. The guanines have lower IP than the adenine in the middle, therefore the adenine acts as a bridge which has to be overcome when CT should occur. Figure 7.8 gives a graphical illustration.

To access the CT characteristics in this sequence, three parameters were calculated: the rate of transfer, the delocalization of the hole and the occupation of the bridge (A). The rate of transfer is calculated by counting the transfers from one guanine to another in a period of 1 ns. Counted are only those transfers, which are not followed by a back transfer within 20 ps. The delocalization of the hole S_{rec} over nucleobases with occupations $|a_i|^2$ is calculated by taking the inverse of the sum of squared occupations.

$$S_{rec} = 1/\sum(|a_i|^2)^2$$

And finally, the population of the bridge is evaluated as a mean value of the occupation of the adenine nucleobase.

Table 7.6: *Comparison of the parametrized CT model with QM/MM simulations of charge transfer in the sequence GAG with the mean-field theory*

Parameter	DNA CT model	QM/MM simulation
rate of transfer (per ns)	2.8 ± 1.1	4.3 ± 2.1
delocalization of the hole	1.31 ± 0.03	1.40 ± 0.34
population of the bridge	0.047 ± 0.003	0.062 ± 0.064

Table 7.6 shows the obtained parameters and the values by T. Kubař ([73], Table 4) are shown as a reference. Comparison of these values shows that the parametrized CT model is able to reproduce the results of QM/MM simulations of charge transfer with the mean-field method in the sequence GAG accurately. All important characteristics are reproduced quantitatively.

While full MD simulations involving the QM/MM scheme need about 2 days to complete one of these simulations, the parametrized CT model needs about 5 min for a simulation of the same length (1 ns).

7.6 Conclusion

In this chapter, the newly developed parametrized model for the CT in DNA was described. The relevant parameters for the CT were fitted to results from QM/MM simulations. Testing was performed on a well-investigated DNA oligomer.

Crucial behaviors and statistics are reproduced quantitatively by the CT model. Therefore, charge transport and charge transfer calculations can be performed with the model. The computational efficiency of the model is about 3 orders of magnitude higher than that of the QM/MM approach.

In conclusion, this CT model makes investigation of very long DNA sequences (up to thousands of base-pairs) over extensive periods of time possible.

CHAPTER **8**

Conclusion

In this work, the charge transport and charge transfer capabilities of DNA under conditions resembling the experimental setup were investigated. Experimental investigations of CT in DNA or other biomolecules are often conducted under conditions, which are not represented accurately in theoretical approaches. First, in common charge transport experiments the DNA strand is attached to electrodes with varying distances. Second, single-molecule experiments (e.g. molecular junctions) are conducted where the solvent is removed partially with a stream of gas. However, the exact amount of water preserved at the DNA strand is unknown and cannot be determined by the experiments alone. Therefore, in this work a simulation setup was developed, which is able to describe these experimental conditions within the limits of computational costs. Based on this setup, calculations of charge transport and charge transfer in stretched an microhydrated DNA were performed. These simulations gave insight into the properties of CT in DNA on an atomistic scale. Hence, the simulations were not only able to explain the experiments, they are also able to predict the outcome of experiments and make suggestions in advance.

As a start, in chapter 4 the structural features of microhydrated DNA and their changes upon stretching stress were investigated with classical MD simulation. It has been shown that both water and counterions are structure-stabilizing factors of DNA. Here, the amount of water and the content of counterions necessary to support a helical double-stranded structure in the simulations was characterized. Upon removal of water from the molecule the helical structure of DNA underwent marked changes. Finally, the double strand lost its helical character roughly at 6 water molecules per phosphate group. Continued drying induced a rupture of hydrogen bonds and a collapse of the double-stranded structure to a disordered compact coil. At fist sight, this was a devastating observation with regard to the experimental setup. Providentially, the simulations showed that the helical structure of microhydrated DNA can be supported by pulling at the 3'-ends, which again resembles the experimental setup.

The further pulling at the 3'-ends of the DNA strands lead to DNA conformations already observed in previous studies. In this work, a crucial dependence on the applied pulling rate was shown for the conformational change from a helix-like to a ladder-like DNA structure. This illustrates the irreversibility of the stretching process in the simulations, when the pulling is several orders of magnitude faster than in experiments. To overcome this issue, the investigation of CT in those stretched DNA conformations was performed on structures held at certain lengths during simulation time, rather than calculating along non-equilibrium pulling simulations.

The first charge transport calculations in this work were performed on DNA in stretched conformations. Here, the efficiency of hole transport in the studied ds-DNA oligomers was little affected by stretching of the molecule by up to 10 % of the free length. When the DNA molecule is stretched further, the response of CT efficiency is determined by the exact nucleobase sequence. In detail, the conformational change lead to a breaking point, where the distance between certain base-pair steps increases largely. Therefore, the overall CT efficiency drops steeply in the exact moment the fist conformational transition occurs.

These results provide a means for the interpretation of the outcome of conductivity experiments where the DNA molecule is stretched and the DNA sequences resemble those simulated. The accumulated knowledge makes it possible to construct DNA sequences such that the extent to which their conductivity is affected by the mechanical stress during the experiments is controlled. Also, suitable DNA sequences can be opted for in the design of DNA-based devices as nano-scale electronic elements, where limited dependence of conductivity on length is desirable.

In the next part, the charge transport and charge transfer in microhydrated DNA strands was investigated. It was shown that the efficiency of charge transfer/transport in microhydrated DNA is similar to that in DNA in bulk solution, provided the regular double-stranded structure is preserved. The magnitude as well as fluctuations of the electric field induced by the aqueous environment are due to the closest hydration shells. The important CT quantities showed rather moderate differences between the fully hydrated and the microhydrated systems. Therefore, it seems conceivable that the transport of electric charge in a microsolvated complex molecular system will exhibit similar characteristics as that in the same system in bulk solution, as long as the structure of the closest solvation shell is retained.

In the final part of this work, a newly developed parametrized model to simulate the CT in DNA was introduced. The relevant parameters for the CT were fitted to results from QM/MM simulations. Testing was performed on a well-investigated DNA oligomer.

Crucial behaviors and statistics are reproduced quantitatively by the CT model. Therefore, charge transport and charge transfer calculations can be performed with the model. The computational efficiency of the model is about 3 orders of magnitude higher than that of the QM/MM approach. In conclusion, this CT model makes investigation of very long DNA sequences (up to thousands of base-pairs) over extensive periods of time possible.

Appendix

DNA Under Experimental Conditions

A.1 The Structural Changes of DNA upon Stretching

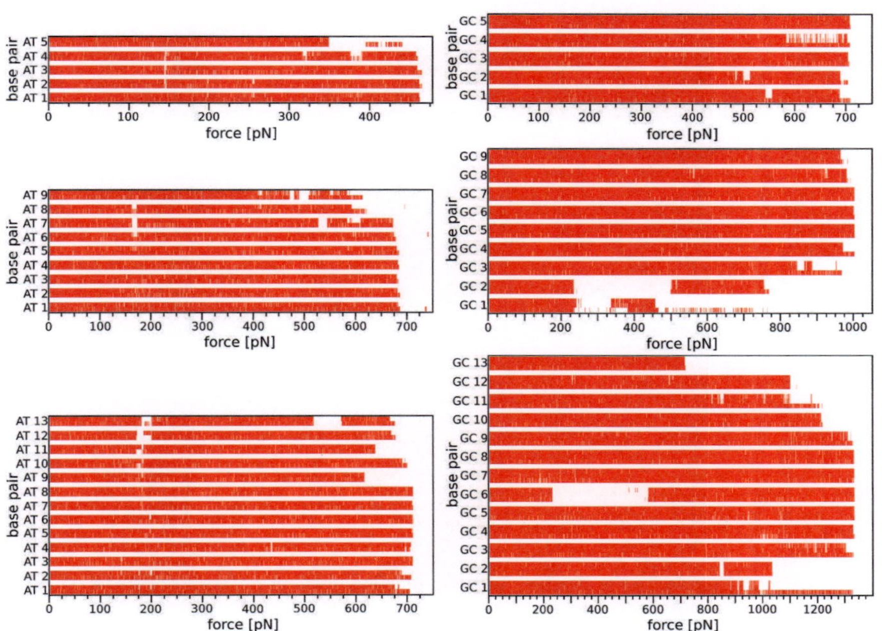

Figure A.1: *Hydrogen-bonding pattern in stretching simulations of fully hydrated DNA oligomers.*

A.2 Effect of Low Hydration and Decreased Ion Content

Hydrogen bond stability in microhydrated DNA

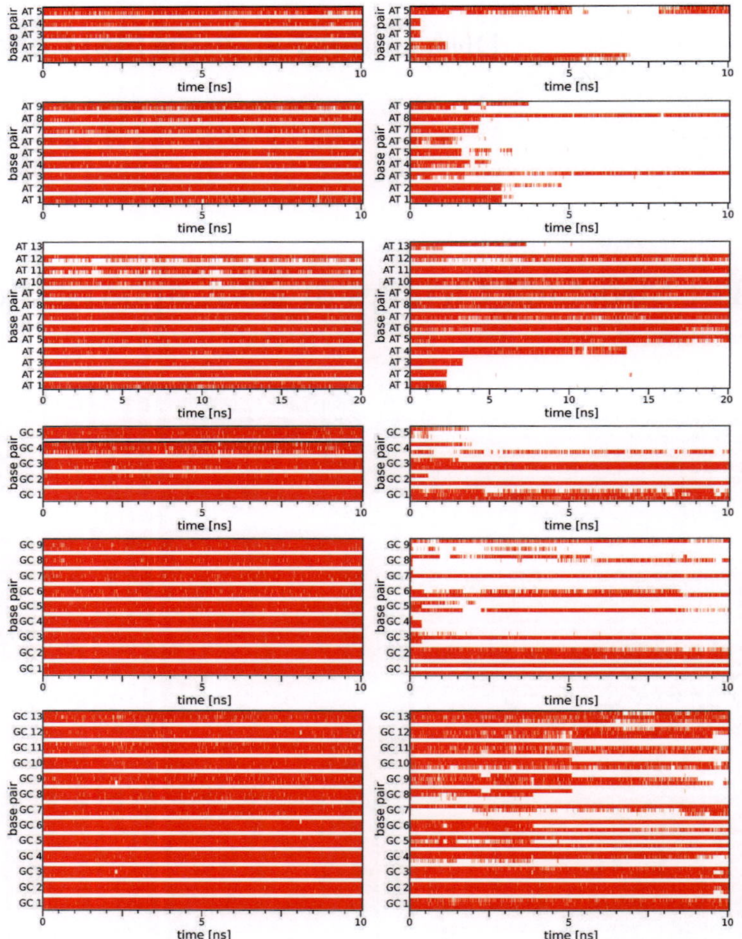

Figure A.2: *The hydrogen-bonding pattern in all oligomers – Dry1 (left) and Dry2 (right).*

Helical Parameters in Microhydrated DNA

Table A.1: Helical parameters of the PolyA DNA species in microhydrated environment in comparison with the values for DNA in bulk solution. Helical parameters were calculated along MD simulations, and mean values with standard deviation are presented for each base-pair step.

species	bp step	rise [nm] complete	rise [nm] Dry1	rise [nm] Dry2	twist [°] complete	twist [°] Dry1	twist [°] Dry2	slide [nm] complete	slide [nm] Dry1	slide [nm] Dry2
A_5	1-2	0.328 ± 0.029	0.326 ± 0.030	0.252 ± 0.058	32.7 ± 6.1	35.3 ± 5.7	-31.6 ± 22.9	-0.047 ± 0.062	0.020 ± 0.039	-0.079 ± 0.102
	2-3	0.339 ± 0.031	0.334 ± 0.029	0.216 ± 0.102	34.2 ± 5.1	34.7 ± 5.8	34.7 ± 14.0	-0.055 ± 0.057	0.010 ± 0.038	0.007 ± 0.096
	3-4	0.339 ± 0.030	0.346 ± 0.028	0.194 ± 0.082	34.1 ± 4.7	39.6 ± 4.4	37.9 ± 30.6	-0.063 ± 0.055	0.003 ± 0.041	0.162 ± 0.060
	4-5	0.333 ± 0.029	0.332 ± 0.024	0.551 ± 0.092	33.2 ± 4.6	36.7 ± 3.6	59.9 ± 32.6	-0.081 ± 0.054	-0.017 ± 0.041	0.222 ± 0.115
A_9	1-2	0.328 ± 0.030	0.328 ± 0.034	0.289 ± 0.085	33.2 ± 6.4	30.1 ± 4.1	30.3 ± 21.7	-0.036 ± 0.063	0.041 ± 0.039	0.114 ± 0.063
	2-3	0.336 ± 0.032	0.349 ± 0.030	0.418 ± 0.087	33.9 ± 5.3	40.4 ± 4.5	61.0 ± 18.1	-0.052 ± 0.059	0.035 ± 0.035	0.161 ± 0.116
	3-4	0.339 ± 0.032	0.321 ± 0.026	0.376 ± 0.033	33.6 ± 5.2	35.6 ± 4.5	60.2 ± 13.3	-0.069 ± 0.057	0.005 ± 0.039	0.217 ± 0.147
	4-5	0.336 ± 0.030	0.325 ± 0.026	0.263 ± 0.044	33.6 ± 4.7	35.0 ± 4.6	11.7 ± 17.5	-0.072 ± 0.056	-0.023 ± 0.050	-0.102 ± 0.130
	5-6	0.334 ± 0.030	0.324 ± 0.027	0.282 ± 0.030	33.7 ± 4.5	37.5 ± 4.8	31.5 ± 5.5	-0.075 ± 0.053	-0.017 ± 0.044	0.102 ± 0.058
	6-7	0.335 ± 0.031	0.321 ± 0.026	0.352 ± 0.050	33.9 ± 4.7	36.0 ± 4.5	58.0 ± 9.3	-0.068 ± 0.055	-0.023 ± 0.041	0.224 ± 0.116
	7-8	0.343 ± 0.031	0.326 ± 0.026	0.280 ± 0.050	33.9 ± 4.6	34.2 ± 5.3	28.0 ± 9.5	-0.070 ± 0.054	-0.013 ± 0.047	0.097 ± 0.046
	8-9	0.329 ± 0.029	0.330 ± 0.027	0.437 ± 0.061	34.6 ± 4.3	38.9 ± 5.0	54.1 ± 14.6	-0.082 ± 0.052	-0.005 ± 0.051	0.110 ± 0.070
A_{13}	1-2	0.328 ± 0.029	0.315 ± 0.028	0.314 ± 0.088	31.8 ± 6.0	29.1 ± 4.8	86.1 ± 76.8	-0.053 ± 0.064	0.044 ± 0.039	0.037 ± 0.219
	2-3	0.336 ± 0.032	0.340 ± 0.029	0.271 ± 0.063	33.6 ± 5.2	39.0 ± 4.8	28.8 ± 15.1	-0.061 ± 0.061	0.036 ± 0.035	-0.048 ± 0.092
	3-4	0.339 ± 0.033	0.325 ± 0.027	0.293 ± 0.087	33.0 ± 5.0	34.5 ± 5.1	38.5 ± 18.7	-0.068 ± 0.059	-0.003 ± 0.042	-0.072 ± 0.111
	4-5	0.337 ± 0.033	0.332 ± 0.027	0.332 ± 0.059	33.3 ± 4.9	37.8 ± 5.0	47.1 ± 9.1	-0.073 ± 0.060	-0.010 ± 0.044	-0.001 ± 0.050
	5-6	0.334 ± 0.031	0.323 ± 0.026	0.324 ± 0.026	32.6 ± 4.5	35.4 ± 4.6	32.5 ± 4.0	-0.079 ± 0.058	-0.020 ± 0.046	0.053 ± 0.057
	6-7	0.335 ± 0.031	0.320 ± 0.026	0.360 ± 0.038	32.8 ± 4.9	36.1 ± 5.2	44.0 ± 8.6	-0.077 ± 0.058	-0.022 ± 0.047	0.084 ± 0.100
	7-8	0.335 ± 0.031	0.316 ± 0.025	0.328 ± 0.027	33.5 ± 4.5	35.4 ± 5.2	34.8 ± 4.7	-0.075 ± 0.058	-0.029 ± 0.048	0.010 ± 0.052
	8-9	0.334 ± 0.030	0.317 ± 0.025	0.345 ± 0.029	32.5 ± 4.4	36.3 ± 4.5	45.2 ± 4.3	-0.084 ± 0.052	-0.032 ± 0.048	0.078 ± 0.049
	9-10	0.335 ± 0.031	0.312 ± 0.025	0.272 ± 0.034	32.9 ± 5.0	34.5 ± 4.6	14.6 ± 4.4	-0.081 ± 0.057	-0.030 ± 0.045	0.040 ± 0.097
	10-11	0.335 ± 0.030	0.326 ± 0.025	0.376 ± 0.031	33.4 ± 4.6	37.0 ± 4.5	43.2 ± 3.5	-0.077 ± 0.057	-0.018 ± 0.044	0.073 ± 0.046
	11-12	0.335 ± 0.030	0.325 ± 0.027	0.365 ± 0.029	32.7 ± 4.6	35.4 ± 3.9	47.5 ± 3.8	-0.081 ± 0.053	-0.015 ± 0.043	0.021 ± 0.047
	12-13	0.330 ± 0.030	0.365 ± 0.035	0.301 ± 0.035	33.6 ± 4.6	56.5 ± 5.0	31.4 ± 5.0	-0.080 ± 0.054	0.101 ± 0.050	0.043 ± 0.065

Table A.2: Helical parameters of the PolyG DNA species in microhydrated environment in comparison with the values for DNA in bulk solution. Helical parameters were calculated along MD simulations, and mean values with standard deviation are presented for each base-pair step.

species	bp step	rise [nm] complete	rise [nm] Dry1	rise [nm] Dry2	twist [°] complete	twist [°] Dry1	twist [°] Dry2	slide [nm] complete	slide [nm] Dry1	slide [nm] Dry2
G_5	1-2	0.338 ± 0.031	0.343 ± 0.031	0.376 ± 0.030	29.3 ± 4.4	32.9 ± 4.0	48.4 ± 6.6	-0.137 ± 0.061	-0.061 ± 0.066	0.096 ± 0.084
	2-3	0.338 ± 0.033	0.335 ± 0.031	0.309 ± 0.035	29.0 ± 4.2	35.4 ± 5.0	22.3 ± 4.9	-0.144 ± 0.056	-0.015 ± 0.067	-0.064 ± 0.088
	3-4	0.339 ± 0.033	0.326 ± 0.031	0.354 ± 0.033	29.1 ± 4.3	34.5 ± 5.0	46.0 ± 5.3	-0.144 ± 0.056	-0.002 ± 0.062	-0.035 ± 0.083
	4-5	0.344 ± 0.034	0.342 ± 0.034	0.283 ± 0.028	29.7 ± 4.5	36.3 ± 5.4	39.4 ± 7.9	-0.142 ± 0.060	-0.029 ± 0.060	-0.040 ± 0.053
G_9	1-2	0.334 ± 0.033	0.317 ± 0.025	0.369 ± 0.042	28.9 ± 4.8	28.0 ± 5.0	29.8 ± 7.3	-0.129 ± 0.069	-0.114 ± 0.058	-0.117 ± 0.066
	2-3	0.336 ± 0.032	0.330 ± 0.026	0.351 ± 0.056	29.1 ± 4.6	29.6 ± 4.0	34.8 ± 6.1	-0.136 ± 0.063	-0.140 ± 0.043	-0.128 ± 0.050
	3-4	0.340 ± 0.031	0.335 ± 0.025	0.343 ± 0.028	29.4 ± 4.1	30.2 ± 3.8	41.5 ± 3.5	-0.145 ± 0.055	-0.153 ± 0.046	-0.005 ± 0.036
	4-5	0.340 ± 0.032	0.333 ± 0.028	0.352 ± 0.026	28.7 ± 4.1	30.3 ± 3.8	33.5 ± 3.7	-0.151 ± 0.053	-0.160 ± 0.051	0.016 ± 0.037
	5-6	0.338 ± 0.031	0.331 ± 0.025	0.288 ± 0.055	28.9 ± 3.9	30.0 ± 4.3	26.9 ± 3.5	-0.148 ± 0.051	-0.165 ± 0.048	0.140 ± 0.040
	6-7	0.339 ± 0.031	0.328 ± 0.026	0.357 ± 0.056	28.9 ± 3.9	28.9 ± 4.0	68.0 ± 30.1	-0.150 ± 0.049	-0.162 ± 0.049	0.226 ± 0.172
	7-8	0.342 ± 0.033	0.330 ± 0.026	0.211 ± 0.052	29.0 ± 4.2	30.9 ± 5.1	45.2 ± 27.0	-0.146 ± 0.055	-0.125 ± 0.065	0.038 ± 0.143
	8-9	0.346 ± 0.033	0.321 ± 0.026	0.233 ± 0.200	29.2 ± 4.2	29.3 ± 4.3	-7.5 ± 23.6	-0.142 ± 0.055	-0.114 ± 0.065	-0.267 ± 0.113
G_{13}	1-2	0.334 ± 0.032	0.325 ± 0.027	0.338 ± 0.059	29.0 ± 4.2	28.5 ± 3.8	40.4 ± 11.7	-0.122 ± 0.063	-0.144 ± 0.048	0.161 ± 0.105
	2-3	0.339 ± 0.033	0.333 ± 0.025	0.261 ± 0.090	29.2 ± 4.2	28.8 ± 4.1	31.5 ± 8.0	-0.142 ± 0.056	-0.154 ± 0.042	0.060 ± 0.071
	3-4	0.342 ± 0.034	0.335 ± 0.025	0.298 ± 0.034	29.2 ± 4.2	29.4 ± 4.9	31.0 ± 5.9	-0.147 ± 0.056	-0.154 ± 0.049	0.143 ± 0.098
	4-5	0.341 ± 0.033	0.332 ± 0.026	0.354 ± 0.034	29.3 ± 4.2	28.7 ± 4.1	46.3 ± 4.0	-0.144 ± 0.056	-0.165 ± 0.051	0.111 ± 0.062
	5-6	0.340 ± 0.032	0.328 ± 0.027	0.312 ± 0.028	29.3 ± 4.5	28.8 ± 4.0	30.8 ± 4.0	-0.140 ± 0.059	-0.173 ± 0.053	0.024 ± 0.056
	6-7	0.341 ± 0.032	0.324 ± 0.025	0.372 ± 0.031	29.2 ± 4.0	28.5 ± 4.6	48.7 ± 3.8	-0.145 ± 0.054	-0.172 ± 0.052	0.114 ± 0.048
	7-8	0.341 ± 0.033	0.321 ± 0.025	0.319 ± 0.024	29.3 ± 4.2	27.4 ± 3.8	31.7 ± 4.5	-0.144 ± 0.057	-0.170 ± 0.044	0.031 ± 0.040
	8-9	0.341 ± 0.032	0.322 ± 0.025	0.324 ± 0.030	29.2 ± 4.1	28.3 ± 4.2	35.1 ± 5.7	-0.144 ± 0.051	-0.171 ± 0.040	0.007 ± 0.048
	9-10	0.338 ± 0.033	0.325 ± 0.026	0.315 ± 0.027	29.0 ± 4.2	29.1 ± 3.9	33.7 ± 4.9	-0.142 ± 0.054	-0.173 ± 0.036	-0.041 ± 0.063
	10-11	0.346 ± 0.033	0.340 ± 0.028	0.324 ± 0.032	29.5 ± 4.1	28.7 ± 3.8	33.1 ± 7.4	-0.145 ± 0.054	-0.157 ± 0.046	-0.083 ± 0.090
	11-12	0.348 ± 0.036	0.345 ± 0.028	0.360 ± 0.043	30.0 ± 4.1	37.6 ± 5.9	32.0 ± 4.7	-0.133 ± 0.059	-0.056 ± 0.056	-0.089 ± 0.089
	12-13	0.343 ± 0.034	0.306 ± 0.028	0.408 ± 0.065	31.4 ± 5.2	26.0 ± 6.4	29.4 ± 5.0	-0.076 ± 0.105	-0.014 ± 0.064	-0.028 ± 0.058

Stabilizing Forces for Microhydrated DNA

Table A.3: *Harmonic force constant (pN/nm) for the Dry1 and Dry2 systems compared to the fully hydrated DNA. Force constant f calculated from the standard deviation of the end-to-end distance σ of the particular DNA oligomer as $f = k_{\mathrm{B}}T/\sigma^2$.*

sequence	full H_2O	Dry1	Dry2
A_5	244	315	990
A_9	225	406	331
A_{13}	84	195	80
G_5	582	187	402
G_9	308	406	72
G_{13}	79	97	515

A.3 Stretching of Microhydrated DNA

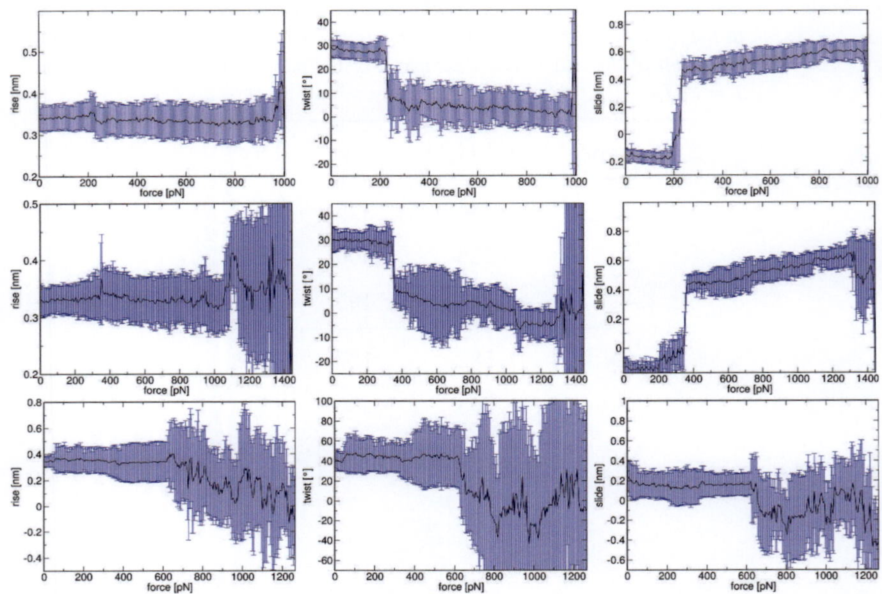

Figure A.3: *Helical parameters of microhydrated* G_9 *(Dry1 – center, Dry2 – bottom) compared to the fully hydrated* G_9 *(top).*

CT in Microhydrated DNA

B.1 Helical Parameters - Complete Overview

Table B.1: *Inter-base-pair helical parameters of all investigated DNA sequences. Mean values for the seven innermost base pairs obtained along 15 ns MD trajectories.*

DNA	Env.	Shift [Å]	Slide [Å]	Rise [Å]	Tilt [°]	Roll [°]	Twist [°]
AA	Full	−0.09	−0.73	3.36	−0.80	0.93	33.20
	Dry1	−0.12	−0.11	3.28	−1.06	6.61	33.80
	Dry2	0.08	0.84	2.93	−1.80	13.00	27.98
AG	Full	−0.09	−0.88	3.38	−0.60	2.90	32.04
	Dry1	−0.31	−0.59	3.34	−1.92	10.46	33.06
	Dry2	−0.58	−1.01	3.42	−3.29	17.70	31.94
AT	Full	0.03	−0.81	3.29	0.00	5.71	30.97
	Dry1	0.07	−0.71	3.23	−0.36	12.77	30.42
	Dry2	−0.62	0.76	2.81	3.18	13.26	39.74
GG	Full	−0.13	−1.47	3.39	−0.51	4.60	29.00
	Dry1	−0.11	−1.53	3.36	−0.66	13.69	29.84
	Dry2	−0.17	−1.55	3.39	−0.40	15.96	31.11
GC	Full	−0.03	−0.40	3.35	−0.13	4.37	32.79
	Dry1	−0.10	0.68	3.34	−0.71	−2.83	37.41
	Dry2	−0.35	1.42	3.43	−3.07	−8.59	39.05
GT	Full	0.09	−0.69	3.34	0.58	5.43	31.65
	Dry1	0.11	−0.81	3.27	0.61	15.46	29.83
	Dry1	0.12	−0.91	3.34	0.61	17.73	31.82

B.2 Electronic Couplings

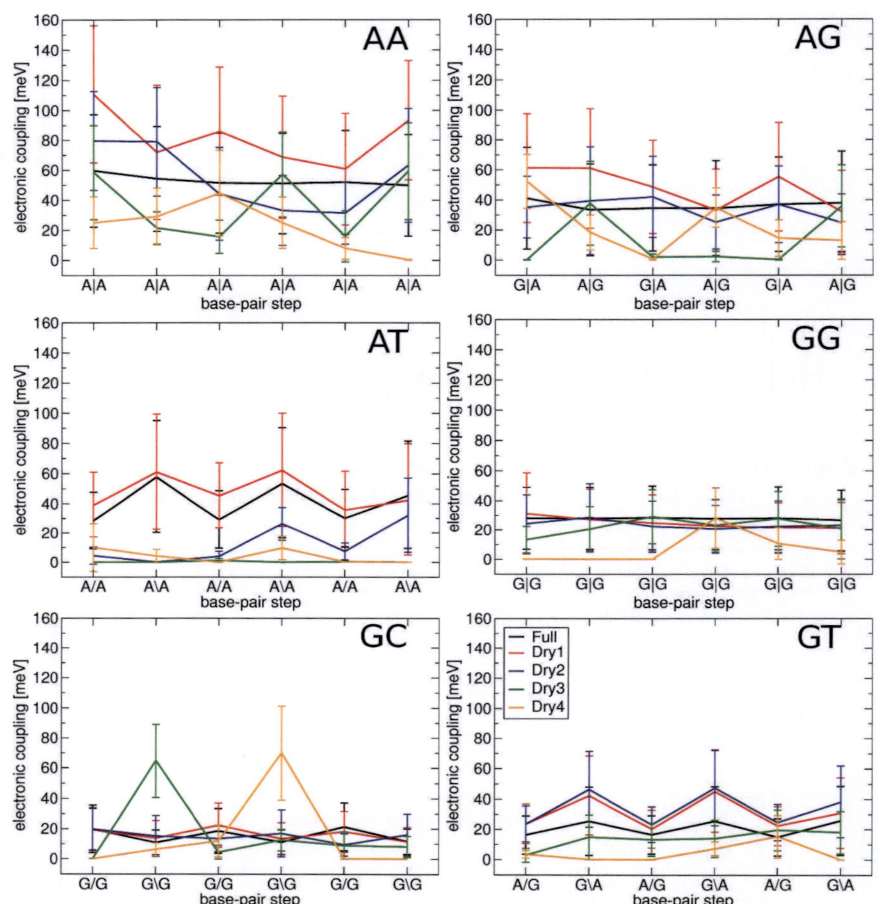

Figure B.1: *Electronic couplings of all investigated DNA sequences in different microhydrated environments. Shown are mean values and std. deviations of T_{ij} of the base-pair steps obtained from 7,500 MD snapshots.*

B.3   Ionization Potentials

Table B.2: *Ionization potentials (mean value ± std. deviation in eV) of the seven innermost adenines for sequence AA in different environments.*

Nucleobase	Full	Dry1	Dry2
A	5.85 ± 0.31	5.37 ± 0.30	5.36 ± 0.32
A	5.87 ± 0.31	5.45 ± 0.29	5.53 ± 0.30
A	5.86 ± 0.31	5.45 ± 0.30	5.49 ± 0.29
A	5.86 ± 0.31	5.44 ± 0.29	5.57 ± 0.30
A	5.86 ± 0.31	5.41 ± 0.30	5.57 ± 0.31
A	5.85 ± 0.31	5.44 ± 0.30	5.59 ± 0.29
A	5.84 ± 0.31	5.49 ± 0.30	5.44 ± 0.29

Table B.3: *Ionization potentials (mean value ± std. deviation in eV) of the seven innermost purines for sequence AG in different environments.*

Nucleobase	Full	Dry1	Dry2
A	5.88 ± 0.31	5.40 ± 0.30	5.58 ± 0.31
G	5.48 ± 0.32	5.03 ± 0.31	5.00 ± 0.32
A	5.86 ± 0.31	5.42 ± 0.30	5.42 ± 0.30
G	5.49 ± 0.32	5.07 ± 0.32	5.10 ± 0.31
A	5.85 ± 0.31	5.48 ± 0.30	5.37 ± 0.29
G	5.47 ± 0.32	5.09 ± 0.31	5.06 ± 0.30
A	5.86 ± 0.31	5.41 ± 0.30	5.30 ± 0.31

Table B.4: *Ionization potentials (mean value $\pm$ std. deviation in eV) of the seven innermost adenines for sequence AT in different environments.*

Nucleobase	Full	Dry1	Dry2
A	5.91 ± 0.31	5.44 ± 0.29	5.55 ± 0.32
A	5.95 ± 0.31	5.52 ± 0.29	5.70 ± 0.31
A	5.96 ± 0.31	5.58 ± 0.29	5.33 ± 0.29
A	5.96 ± 0.31	5.59 ± 0.29	5.31 ± 0.28
A	5.98 ± 0.30	5.54 ± 0.29	5.32 ± 0.27
A	5.96 ± 0.31	5.57 ± 0.30	5.36 ± 0.29
A	5.93 ± 0.31	5.47 ± 0.29	5.27 ± 0.29

Table B.5: *Ionization potentials (mean value $\pm$ std. deviation in eV) of the eight innermost purines for sequence DD in different environments.*

Nucleobase	Full	Dry1	Dry2
G	5.45 ± 0.32	4.99 ± 0.31	5.14 ± 0.32
G	5.43 ± 0.32	4.99 ± 0.31	5.14 ± 0.32
A	5.82 ± 0.31	5.44 ± 0.30	5.55 ± 0.32
A	5.91 ± 0.30	5.55 ± 0.29	5.56 ± 0.31
A	5.92 ± 0.30	5.54 ± 0.29	5.49 ± 0.29
A	5.83 ± 0.31	5.41 ± 0.29	5.55 ± 0.30
G	5.44 ± 0.32	4.97 ± 0.30	4.99 ± 0.30
G	5.44 ± 0.32	5.00 ± 0.31	5.08 ± 0.30

Table B.6: *Ionization potentials (mean value $\pm$ std. deviation in eV) of the seven innermost guanines for sequence GG in different environments.*

Nucleobase	Full	Dry1	Dry2
G	5.50 ± 0.33	5.08 ± 0.32	5.10 ± 0.31
G	5.50 ± 0.32	5.12 ± 0.32	5.12 ± 0.31
G	5.50 ± 0.32	5.13 ± 0.31	5.10 ± 0.31
G	5.50 ± 0.32	5.10 ± 0.31	5.04 ± 0.31
G	5.50 ± 0.32	5.09 ± 0.31	5.01 ± 0.31
G	5.50 ± 0.32	5.07 ± 0.31	5.01 ± 0.31
G	5.49 ± 0.32	5.08 ± 0.32	5.05 ± 0.31

Table B.7: *Ionization potentials (mean value $\pm$ std. deviation in eV) of the seven innermost guanines for sequence GC in different environments.*

Nucleobase	Full	Dry1	Dry2
G	5.44 ± 0.44	5.10 ± 0.30	5.02 ± 0.29
G	5.42 ± 0.44	5.03 ± 0.31	5.00 ± 0.31
G	5.42 ± 0.41	5.10 ± 0.31	5.08 ± 0.31
G	5.42 ± 0.40	4.96 ± 0.30	4.93 ± 0.30
G	5.41 ± 0.41	5.08 ± 0.31	5.03 ± 0.30
G	5.41 ± 0.40	4.94 ± 0.31	4.90 ± 0.30
G	5.42 ± 0.43	5.03 ± 0.31	5.13 ± 0.30

B.4 ESP Induced by Different Groups of Atoms

Table B.8: ESP (in V) induced by the different groups of atoms, in simulations of DNA in different environments. Mean values ($\pm$ std. deviations) over 15 ns MD trajectories.

DNA	Env.	System	DNA	Water	Na$^+$
AA	Full	0.49 ± 0.29	-25.31	7.76	18.05
	Dry1	0.10 ± 0.28	-26.61	1.01	25.70
	Dry2	0.20 ± 0.30	-28.41	0.33	28.28
AG	Full	0.52 ± 0.30	-22.19	6.70	16.01
	Dry1	0.10 ± 0.30	-23.81	0.60	23.31
	Dry2	0.10 ± 0.29	-25.10	0.67	24.52
AT	Full	0.47 ± 0.29	-22.29	7.29	15.47
	Dry1	0.04 ± 0.28	-23.80	1.07	22.77
	Dry2	-0.01 ± 0.31	-25.85	0.28	25.56
GG	Full	0.61 ± 0.31	-24.26	6.16	18.71
	Dry1	0.25 ± 0.30	-26.48	0.37	26.37
	Dry2	0.21 ± 0.30	-27.84	0.54	27.50
GC	Full	0.51 ± 0.31	-22.60	6.45	16.67
	Dry1	0.11 ± 0.30	-24.04	0.75	23.40
	Dry2	0.17 ± 0.30	-24.49	0.45	24.20
GT	Full	0.50 ± 0.30	-25.25	6.91	18.85
	Dry1	0.11 ± 0.29	-26.91	0.76	26.25
	Dry2	0.10 ± 0.29	-28.19	0.84	27.45

B.5 Distance of Charged Atom Groups from the Helical Axis

Table B.9: *Distance (in nm) of phosphate groups in the DNA backbones and of Na$^+$ counterions from the helical axis of DNA, in simulations of DNA in different environments.*

		Na$^+$	Phosphate
AA	Full	1.491	1.021
	Dry1	0.999	0.980
	Dry2	0.883	0.892
AG	Full	1.448	1.047
	Dry1	0.962	1.000
	Dry2	0.785	0.892
AT	Full	1.513	1.069
	Dry1	1.017	1.022
	Dry2	0.912	0.947
GG	Full	1.468	1.142
	Dry1	0.940	1.025
	Dry2	0.799	0.910
GC	Full	1.423	1.007
	Dry1	0.929	0.896
	Dry2	0.841	0.817
GT	Full	1.454	1.048
	Dry1	0.968	1.012
	Dry2	0.811	0.893

[1] N. A. Zimbovskaya and M. R. Pederson. Electron transport through molecular junctions. *Phys. Rev.*, **2011**. 509, 1, 1–87.

[2] L. D. A. Siebbeles and F. C. Grozema, editors. *Charge and Exciton Transport through Molecular Wires*. Wiley, New York, **2011**.

[3] H.-W. Fink and C. Schönenberger. Electrical conduction through DNA molecules. *Nature*, **1999**. 398, 6726, 407–410.

[4] K.-H. Yoo, D. H. Ha, J.-O. Lee, J. W. Park, J. Kim, J. J. Kim, H.-Y. Lee, T. Kawai, and H. Y. Choi. Electrical conduction through poly(dA)-poly(dT) and poly(dG)-poly(dC) DNA molecules. *Phys. Rev. Lett.*, **2001**. 87, 19, 198102.

[5] A. Y. Kasumov, M. Kociak, S. Guéron, B. Reulet, V. T. Volkov, D. V. Klinov, and H. Bouchiat. Proximity-Induced Superconductivity in DNA. *Science*, **2001**. 291, 5502, 280–282.

[6] A. J. Storm, J. van Noort, S. de Vries, and C. Dekker. Insulating behavior for DNA molecules between nanoelectrodes at the 100 nm length scale. *Appl. Phys. Lett.*, **2001**. 79, 23, 3881–3883.

[7] D. Porath, A. Bezryadin, S. De Vries, and C. Dekker. Direct measurement of electrical transport through DNA molecules. *Nature*, **2000**. 403, 635–638.

[8] B. Xu, P. Zhang, X. Li, and N. Tao. Direct conductance measurement of single DNA molecules in aqueous solution. *Nano Lett.*, **2004**. 4, 1105–1108.

[9] H. Cohen, C. Nogues, R. Naaman, and D. Porath. Direct measurement of electrical transport through single DNA molecules of complex sequence. *Proc. Natl. Acad. Sci. USA*, **2005**. 102, 11589–11593.

[10] J. Hihath, B. Q. Xu, P. M. Zhang, and N. J. Tao. Study of single-nucleotide

polymorphisms by means of electrical conductance measurements. *Proc. Natl. Acad. Sci. USA*, **2005**. 102, 47, 16979–16983.

[11] H. van Zalinge, D. J. Schiffrin, A. D. Bates, W. Haiss, J. Ulstrup, and R. J. Nichols. Single-Molecule Conductance Measurements of Single- and Double-Stranded DNA Oligonucleotides. *ChemPhysChem*, **2006**. 7, 94–98.

[12] N. Kang, A. Erbe, and E. Scheer. Electrical characterization of DNA in mechanically controlled break-junctions. *New J. Phys.*, **2008**. 10, 023030.

[13] X. Guo, A. A. Gorodetsky, J. Hone, J. K. Barton, and C. Nuckolls. Conductivity of a single DNA duplex bridging a carbon nanotube gap. *Nature Nanotechnol.*, **2008**. 3, 3, 163–167.

[14] C. Joachim and M. A. Ratner. Molecular electronics: Some views on transport junctions and beyond. *Proc. Natl. Acad. Sci. USA*, **2005**. 102, 8801–8808.

[15] G. Cuniberti, G. Fagas, and K. Richter, editors. *Introducing Molecular Electronics*, volume 680 of *Lecture Notes in Physics*. Springer, Berlin, **2005**.

[16] M. H. F. Wilkins, A. R. Stokes, and H. R. Wilson. Molecular structure of nucleic acids: Molecular structure of deoxypentose nucleic acids. *Nature*, **1953**. 171, 4356, 738–740.

[17] J. D. Watson and F. H. C. Crick. Molecular structure of nucleic acids: A structure for deoxyribose nucleic acid. *Nature*, **1953**. 171, 4356, 737–738.

[18] C. Bustamante, J. F. Marko, E. D. Siggia, and S. Smith. Entropic elasticity of λ-phage DNA. *Science*, **1994**. 265, 5178, 1599–1600.

[19] P. Cluzel, A. Lebrun, C. Heller, R. Lavery, J.-L. Viovy, D. Chatenay, and F. Caron. DNA: An extensible molecule. *Science*, **1996**. 271, 5250, 792–794.

[20] S. B. Smith, Y. Cui, and C. Bustamante. Overstretching B-DNA: The elastic response of individual double-stranded and single-stranded DNA molecules. *Science*, **1996**. 271, 5250, 795–799.

[21] A. Lebrun and R. Lavery. Modelling extreme stretching of DNA. *Nucleic Acids Res.*, **1996**. 24, 12, 2260–2267.

[22] M. W. Konrad and J. I. Bolonick. Molecular dynamics simulation of DNA stretching is consistent with the tension observed for extension and strand

separation and predicts a novel ladder structure. *J. Am. Chem. Soc.*, **1996**. 118, 45, 10989–10994.

[23] J. F. Allemand, D. Bensimon, R. Lavery, and V. Croquette. Stretched and overwound DNA forms a Pauling-like structure with exposed bases. *Proc. Natl. Acad. Sci. USA*, **1998**. 95, 24, 14152–14157.

[24] C. Bustamante, S. B. Smith, J. Liphardt, and D. Smith. Single-molecule studies of DNA mechanics. *Curr. Opin. Struct. Biol.*, **2000**. 10, 3, 279–285.

[25] H. Clausen-Schaumann, M. Seitz, R. Krautbauer, and H. E. Gaub. Force spectroscopy with single bio-molecules. *Curr. Opin. Chem. Biol.*, **2000**. 4, 5, 524–530.

[26] I. Rouzina and V. A. Bloomfield. Force-induced melting of the DNA double helix 1. Thermodynamic analysis. *Biophys. J.*, **2001**. 80, 2, 882–893.

[27] S. Cocco, J. Yan, J.-F. Léger, D. Chatenay, and J. F. Marko. Overstretching and force-driven strand separation of double-helix DNA. *Phys. Rev. E*, **2004**. 70, 1, 011910.

[28] J. van Mameren, P. Gross, G. Farge, P. Hooijman, M. Modesti, M. Falkenberg, G. J. L. Wuite, and E. J. G. Peterman. Unraveling the structure of DNA during overstretching by using multicolor, single-molecule fluorescence imaging. *Proc. Natl. Acad. Sci. USA*, **2009**. 106, 43, 18231–18236.

[29] C. H. Albrecht, G. Neuert, R. A. Lugmaier, and H. E. Gaub. Molecular force balance measurements reveal that double-stranded DNA unbinds under force in rate-dependent pathways. *Biophys. J.*, **2008**. 94, 12, 4766–4774.

[30] C. Danilowicz, C. Limouse, K. Hatch, A. Conover, V. W. Coljee, N. Kleckner, and M. Prentiss. The structure of DNA overstretched from the 5ˇo325ˇo32 ends differs from the structure of DNA overstretched from the 3ˇo323ˇo32 ends. *Proc. Natl. Acad. Sci. USA*, **2009**. 106, 32, 13196–13201.

[31] H. Fu, H. Chen, J. F. Marko, and J. Yan. Two distinct overstretched DNA states. *Nucleic Acids Res.*, **2010**. 38, 16, 5594–5600.

[32] K. R. Chaurasiya, T. Paramanathan, M. J. McCauley, and M. C. Williams. Biophysical characterization of DNA binding from single molecule force mea-

surements. *Phys. Life Rev.*, **2010**. 7, 3, 299 and comments following in the same issue.

[33] K. M. Kosikov, A. A. Gorin, V. B. Zhurkin, and W. K. Olson. DNA stretching and compression: Large-scale simulations of double helical structures. *J. Mol. Biol.*, **1999**. 289, 5, 1301–1326.

[34] S. Piana. Structure and energy of a DNA dodecamer under tensile load. *Nucleic Acids Res.*, **2005**. 33, 22, 7029–7038.

[35] R. Lohikoski, J. Timonen, and A. Laaksonen. Molecular dynamics simulation of single DNA stretching reveals a novel structure. *Chem. Phys. Lett.*, **2005**. 407, 1-3, 23–29.

[36] D. R. Roe and A. M. Chaka. Structural basis of pathway-dependent force profiles in stretched DNA. *J. Phys. Chem. B*, **2009**. 113, 46, 15364–15371.

[37] J. Řezáč, P. Hobza, and S. A. Harris. Stretched DNA investigated using molecular-dynamics and quantum-mechanical calculations. *Biophys. J.*, **2010**. 98, 1, 101–110.

[38] P. Maragakis, R. L. Barnett, E. Kaxiras, M. Elstner, and T. Frauenheim. Electronic structure of overstretched DNA. *Phys. Rev. B*, **2002**. 66, 24, 241104.

[39] B. Song, M. Elstner, and G. Cuniberti. Anomalous Conductance Response of DNA Wires under Stretching. *Nano Lett.*, **2008**. 8, 10, 3217–3220.

[40] M. A. O'Neill, H.-C. Becker, C. Z. Wan, J. K. Barton, and A. H. Zewail. Ultrafast dynamics in DNA-mediated electron transfer: Base gating and the role of temperature. *Angew. Chem. Int. Ed.*, **2003**. 42, 47, 5896–5900.

[41] S. S. Skourtis, D. H. Waldeck, and D. N. Beratan. Fluctuations in Biological and Bioinspired Electron-Transfer Reactions. *Annu. Rev. Phys. Chem.*, **2010**. 61, 461–485.

[42] D. Q. Andrews, R. P. Van Duyne, and M. A. Ratner. Stochastic modulation in molecular electronic transport junctions: Molecular dynamics coupled with charge transport calculations. *Nano Lett.*, **2008**. 8, 4, 1120–1126.

[43] R. Maul and W. Wenzel. Influence of structural disorder and large-scale

geometric fluctuations on the coherent transport of metallic junctions and molecular wires. *Phys. Rev. B*, **2009**. 80, 4, 045424.

[44] P. B. Woiczikowski, T. Kubař, R. Gutiérrez, R. A. Caetano, G. Cuniberti, and M. Elstner. Combined density functional theory and Landauer approach for hole transfer in DNA along classical molecular dynamics trajectories. *J. Chem. Phys.*, **2009**. 130, 21, 215104.

[45] R. Gutiérrez, R. A. Caetano, P. B. Woiczikowski, T. Kubař, M. Elstner, and G. Cuniberti. Charge Transport through Biomolecular Wires in a Solvent: Bridging Molecular Dynamics and Model Hamiltonian Approaches. *Phys. Rev. Lett.*, **2009**. 102, 20, 208102.

[46] P. Makk, D. Visontai, L. Oroszlany, D. Z. Manrique, S. Csonka, J. Cserti, C. Lambert, and A. Halbritter. Advanced Simulation of Conductance Histograms Validated through Channel-Sensitive Experiments on Indium Nanojunctions. *Phys. Rev. Lett.*, **2011**. 107, 27, 276801.

[47] R. E. Franklin and R. G. Gosling. The Structure of Sodium Thymonucleate Fibres I. The Influence of Water Content. *Acta Cryst.*, **1953**. 6, 8-9, 673–677.

[48] A. K. Mazur. DNA dynamics in a water drop without counterions. *J. Am. Chem. Soc.*, **2002**. 124, 49, 14707–14715.

[49] A. K. Mazur. Titration in silico of reversible B ž194 A transitions in DNA. *J. Am. Chem. Soc.*, **2003**. 125, 26, 7849–7859.

[50] C. I. Bayly, P. Cieplak, W. D. Cornell, and P. A. Kollman. A Well-Behaved Electrostatic Potential Based Method Using Charge Restraints for Deriving Atomic Charges: The RESP Model. *J. Phys. Chem.*, **1993**. 97, 10269–10280.

[51] P. P. Ewald. Die Berechnung optischer und elektrostatischer Gitterpotentiale. *Annalen der Physik*, **1921**. 369, 3, 253–287.

[52] U. Essmann, L. Perera, M. L. Berkowitz, T. Darden, H. Lee, and L. G. Pedersen. Particle mesh Ewald - an n.log(n) method for Ewald sums in large systems. *J. Chem. Phys.*, **1993**. 98, 12, 10089–10092.

[53] T. Darden, D. York, and L. G. Pedersen. A smooth particle mesh Ewald method. *J. Chem. Phys.*, **1995**. 103, 19, 8577.

[54] I. Newton. *Philosophiæ naturalis principia mathematica*, **1687**. London.

[55] R. W. Hockney, S. P. Goel, and J. W. Eastwood. Quiet high-resolution computer models of a plasma. *J. Chem. Phys.*, **1974**. 14, 2, 148–158.

[56] L. Verlet. Computer "Experiments" on Classical Fluids. I. Thermodynamical Properties of Lennard-Jones Molecules. *Phys. Rev.*, **1967**. 159, 98.

[57] H. J. C. Berendsen, J. P. M. Postma, W. F. van Gunsteren, A. DiNola, and J. R. Haak. Molecular-Dynamics with Coupling to an External Bath. *J. Chem. Phys.*, **1984**. 81, 8, 3684–3690.

[58] S. Nosé. A unified formulation of the constant temperature molecular-dynamics methods. *J. Chem. Phys.*, **1984**. 81, 1, 511–519.

[59] W. G. Hoover. Canonical dynamics: Equilibrium phase-space distributions. *Phys. Rev. A*, **1985**. 31, 3, 1695–1697.

[60] S. Nosé and M. L. Klein. Constant pressure molecular dynamics for molecular systems. *Mol. Phys.*, **1983**. 50, 5, 1055–1076.

[61] M. Parinello and A. Rahman. Polymorphic transitions in single crystals: A new molecular dynamics method. *J. Appl. Phys.*, **1981**. 52, 7182–7190.

[62] P. Hohenberg and W. Kohn. Inhomogeneous Electron Gas. *Phys. Rev.*, **1964**. 136, 864–871.

[63] M. Born and R. Oppenheimer. Zur Quantentheorie der Molekeln. *Annalen der Physik*, **1927**. 84, 457–484.

[64] P. Hohenberg and L. J. Sham. Self-Consistent Equations Inclouding Exchange and Correlation Effects. *Phys. Rev.*, **1965**. 140, 1133–1138.

[65] M. Elstner. Semi-empirical DFT: SCC-DFTB, **2008**. Lecture Script.

[66] M. Elstner. The SCC-DFTB method and its application to biological systems. *Theoret. Chem. Acc.*, **2006**. 116, 316–325.

[67] M. Elstner, D. Porezag, G. Jungnickel, J. Elsner, M. Haugk, T. Frauenheim, S. Suhai, and G. Seifert. Self-consistent-charge density-functional tightbinding method for simulations of complex materials properties. *Phys. Rev. B*, **1998**. 58, 7260–7268.

[68] M. Elstner. SCC-DFTB: What is the proper degree of self-consistency. *Phys. Rev. B*, **2007**. 111, 5614–5621.

[69] R. Gutiérrez, R. Caetano, P. B. Woiczikowski, T. Kubař, M. Elstner, and G. Cuniberti. Structural fluctuations and quantum transport through DNA molecular wires: a combined molecular dynamics and model Hamiltonian approach. *New J. Phys.*, **2010**. 12, 023022.

[70] P. B. Woiczikowski, T. Kubař, R. Gutiérrez, G. Cuniberti, and M. Elstner. Structural stability versus conformational sampling in biomolecular systems: Why is the charge transfer efficiency in G4-DNA better than in double-stranded DNA? *J. Chem. Phys.*, **2010**. 133, 3, 035103.

[71] R. A. Marcus. On the theory of oxidation-reduction reactions involving electron transfer. I. *J. Chem. Phys.*, **1956**. 24, 5, 966–978.

[72] T. Kubař and M. Elstner. A hybrid approach to simulation of electron transfer in complex molecular systems. *J. R. Soc. Interface*, **2013**. page to be published.

[73] T. Kubař and M. Elstner. Efficient algorithms for the simulation of non-adiabatic electron transfer in complex molecular systems: application to DNA. *Phys. Chem. Chem. Phys.*, **2013**. 15, 5794–5813.

[74] E. Hückel. Quantentheoretische Beiträge zum Benzolproblem. *Z. Phys.*, **1931**. 70, 204–286.

[75] E. Hückel. Quantentheoretische Beiträge zum Problem der aromatischen und ungesättigten Verbindungen. III. *Z. Phys.*, **1932**. 76, 628–648.

[76] K. Kitaura, E. Ikeo, T. Asada, T. Nakano, and U. M. Fragment molecular orbital method: an approximate computational method for large molecules. *Chem. Phys. Lett.*, **1999**. 313, 701–706.

[77] M. Bixon and J. Jortner. Charge Transport in DNA Via Thermally Induced Hopping. *J. Am. Chem. Soc.*, **2001**. 123, 50, 12556–12567.

[78] R. G. Endres, D. L. Cox, and R. R. P. Singh. Colloquium: The quest for high-conductance DNA. *Rev. Mod. Phys.*, **2004**. 76, 1, 195–214.

[79] T. Kubař and M. Elstner. Coarse-Grained Time-Dependent Density Func-

tional Simulation of Charge Transfer in Complex Systems: Application to Hole Transfer in DNA. *J. Phys. Chem. B*, **2010**. 114, 34, 11221–11240.

[80] J. Blumberger and G. Lamoureux. Reorganization free energies and quantum corrections for a model electron self-exchange reaction: Comparison of polarizable and non-polarizable solvent models. *Mol. Phys.*, **2008**. 106, 1597–1611.

[81] E. Hatcher, A. Balaeff, S. Keinan, R. Venkatramani, and D. N. Beratan. PNA versus DNA: Effects of structural fluctuations on electronic structure and hole-transport mechanisms. *J. Am. Chem. Soc.*, **2008**. 130, 35, 11752–11761.

[82] R. Venkatramani, S. Keinan, A. Balaeff, and D. N. Beratan. Nucleic acid charge transfer: Black, white and gray. *Coord. Chem. Rev.*, **2011**. 255, 7-8, SI, 635–648.

[83] D. S. Fisher and P. A. Lee. Relation Between Conductivity and Transmission Matrix. *Phys. Rev. B*, **1981**. 23, 12, 6851–6854.

[84] R. Venkatramani, K. L. Davis, E. Wierzbinski, S. Bezer, A. Balaeff, S. Keinan, A. Paul, L. Kocsis, D. N. Beratan, C. Achim, and D. H. Waldeck. Evidence for a Near-Resonant Charge Transfer Mechanism for Double-Stranded Peptide Nucleic Acid. *J. Am. Chem. Soc.*, **2011**. 133, 1, 62–72.

[85] R. A. Marcus and N. Sutin. Electron transfers in chemistry and biology. *Biochimica et Biophysica Acta*, **1985**. 811, 3, 265–322.

[86] V. G. Levich and R. R. Dogonadze. Theory of non-radiation electron transitions from ion to ion in solutions. *Dokl. Akad. Nauk SSSR*, **1959**. 124, 123–126.

[87] N. S. Hush. Adiabatic theory of outer sphere electron-transfer reactions in solution. *Trans. Faraday Soc.*, **1961**. 57, 557–580.

[88] J. J. Hopfield. Electron-transfer between biological molecules by thermally activated tunneling. *Proc. Natl. Acad. Sci. USA*, **1974**. 71, 3640–3644.

[89] J. Jortner. Temperature-dependent activation-energy for electron-transfer between biological molecules. *J. Chem. Phys.*, **1976**. 64, 4860–4867.

[90] D. W. Small, D. V. Matyushov, and G. A. Voth. The theory of electron transfer reactions: What may be missing? *J. Am. Chem. Soc.*, **2003**. 125, 7470–7478.

[91] T. Kubař, P. B. Woiczikowski, G. Cuniberti, and M. Elstner. Efficient calcu-

lation of charge-transfer matrix elements for hole transfer in DNA. *J. Phys. Chem. B*, **2008**. 112, 26, 7937–7947.

[92] B. J. A. and T. Chakraborty. Energetics of the hole transfer in DNA duplex oligomers. *Chem. Phys. Lett.*, **2007**. 446, 159.

[93] T. Kubař and M. Elstner. Solvent reorganization energy of hole transfer in DNA. *J. Phys. Chem. B*, **2009**. 113, 16, 5653–5656.

[94] P. Ehrenfest. Bemerkung über die angenäherte Gültigkeit der klassischen Mechanik innerhalb der Quantenmechanik. *Z. Phys. A*, **1927**. 45, 455–457.

[95] A. D. McLachlan. A variational solution of the time-dependent Schrödinger equation. *Mol. Phys.*, **1964**. 8, 39–44.

[96] P. T. Henderson, D. Jones, G. Hampikian, Y. Kan, and S. G. B. Long-distance charge transport in duplex DNA: The phonon-assisted polaron-like hopping mechanism. *Proc. Natl. Acad. Sci. USA*, **1999**. 96, 8353–8358.

[97] E. M. Conwell and S. V. Rakhmanova. Polarons in DNA. *Proc. Natl. Acad. Sci. USA*, **2000**. 97, 4556–4560.

[98] C.-S. Liu and S. G. B. Base sequence effects in radical cation migration in duplex DNA: Support for the polaron-like hopping model. *J. Am. Chem. Soc.*, **2003**. 125, 6098–6102.

[99] E. M. Conwell. Charge transport in DNA in solution: The role of polarons. *Proc. Natl. Acad. Sci. USA*, **2005**. 102, 8795–8799.

[100] F. C. Grozema, S. Tonzani, Y. A. Berlin, G. C. Schatz, L. D. A. Siebbeles, and M. A. Ratner. Effect of Structural Dynamics on Charge Transfer in DNA Hairpins. *J. Am. Chem. Soc.*, **2008**. 130, 15, 5157–5166.

[101] E. Hatcher, A. Balaeff, R. Keinan S, Venkatramani, and D. N. Beratan. PNA versus DNA: effects of structural fluctuations on electronic structure and hole-transport mechanisms. *J. Am. Chem. Soc.*, **2008**. 130, 35, 11752–11761.

[102] T. Kubař, U. Kleinekathöfer, and M. Elstner. Solvent fluctuations drive the hole transfer in DNA: a mixed quantum–classical study. *J. Phys. Chem. B*, **2009**. 113, 13107–13117.

[103] J. C. Tully and R. K. Preston. Trajectory surface hopping approach to nonadi-

abatic molecular collisions: The reaction of H^+ with D_2. *J. Chem. Phys.*, **1971**. 55, 1061–1071.

[104] G. Granucci, M. Persico, and A. Toniolo. Direct semiclassical simulation of photochemical processes with semiempirical wave functions. *J. Chem. Phys.*, **2001**. 114, 10608–10615.

[105] W. L. Peticolas, Y. Wang, and G. A. Thomas. Some Rules for Predicting the Base-Sequence Dependence of DNA Conformation. *Proc. Natl. Acad. Sci. USA*, **1988**. 85, 2579–2583.

[106] R. E. Dickerson et al. Definitions and nomenclature of nucleic acid structure components. *Nucleic Acids Res.*, **1989**. 17, 5, 1797–1803.

[107] X.-J. Lu and W. K. Olson. 3DNA: a software package for the analysis, rebuilding and visualization of three-dimensional nucleic acid structures. *Nucleic Acids Res.*, **2003**. 31, 5108–5121.

[108] H. R. Drew, R. M. Wing, T. Takano, C. Broka, S. Tanaka, K. Itakura, and R. E. Dickerson. Structure of a B-DNA dodecamer: conformation and dynamics. *Proc. Natl. Acad. Sci. USA*, **1981**. 78, 4, 2179–2183.

[109] J. Stroud. make-na server, **2004**. Http://structure.usc.edu/make-na/server.html.

[110] W. L. Jorgensen, J. Chandrasekhar, J. D. Madura, R. W. Impey, and M. L. Klein. Comparison of simple potential functions for simulating liquid water. *J. Chem. Phys.*, **1983**. 79, 2, 926–935.

[111] W. D. Cornell, P. Cieplak, C. I. Bayly, I. R. Gould, K. M. Merz, D. M. Ferguson, D. C. Spellmeyer, T. Fox, J. W. Caldwell, and P. A. Kollman. A 2nd generation force-field for the simulation of proteins, nucleic-acids, and organic-molecules. *J. Am. Chem. Soc.*, **1995**. 117, 19, 5179–5197.

[112] J. Wang, P. Cieplak, and P. A. Kollman. How Well Does a Restrained Electrostatic Potential (RESP) Model Perform in Calculating Conformational Energies of Organic and Biological Molecules? *J. Comput. Chem.*, **2000**. 21, 1049–1074.

[113] A. Pérez, I. Marchán, D. Svozil, J. Šponer, T. E. Cheatham III, C. A. Laughton,

and M. Orozco. Refinement of the AMBER force field for nucleic acids: Improving the description of α/γ conformers. *Biophys. J.*, **2007**. 92, 3817–3829.

[114] J. Åqvist. Ion–Water Interaction Potentials Derived from Free-Energy Perturbation Simulations. *J. Phys. Chem.*, **1990**. 94, 21, 8021–8024.

[115] AmberTools 1.4, **2010**. Http://ambermd.org.

[116] F. Ryjáček. Ambconv, **2002**. Http://www.gromacs.org/ Downloads/User_contributions/Other_software.

[117] B. Hess, H. Bekker, H. J. C. Berendsen, and J. G. E. M. Fraaije. LINCS: A linear constraint solver for molecular simulations. *J. Comput. Chem.*, **1997**. 18, 12, 1463–1472.

[118] D. van der Spoel, E. Lindahl, B. Hess, G. Groenhof, A. E. Mark, and H. J. C. Berendsen. GROMACS: Fast, flexible, and free. *J. Comput. Chem.*, **2005**. 26, 1701–1718.

[119] B. Hess, C. Kutzner, D. van der Spoel, and E. Lindahl. GROMACS 4: Algorithms for highly efficient, load-balanced, and scalable molecular simulation. *J. Chem. Theory Comput.*, **2008**. 4, 3, 435–447.

[120] P. Hobza, M. Kabeláč, J. Šponer, P. Mejzlík, and J. Vondrášek. Performance of empirical potentials AMBER, CFF95, CVFF, CHARMM, OPLS, POLTEV), semiempirical quantum chemical methods (AM1, MNDO/M, PM3), and ab initio Hartree–Fock method for interaction of DNA bases: Comparison with nonempirical beyond Hartree–Fock results. *J. Comput. Chem.*, **1997**. 18, 9, 1136–1150.

[121] M. Zgarbová, M. Otyepka, J. Šponer, P. Hobza, and P. Jurečka. Large-scale compensation of errors in pairwise-additive empirical force fields: comparison of AMBER intermolecular terms with rigorous DFT-SAPT calculations. *Phys. Chem. Chem. Phys.*, **2010**. 12, 35, 10476–10493.

[122] M. Rueda, S. G. Kalko, F. J. Luque, and M. Orozco. The Structure and Dynamics of DNA in the Gas Phase. *J. Am. Chem. Soc.*, **2003**. 125, 26, 8007–8014.

[123] A. Voityuk. Fluctuation of the electronic coupling in DNA: Multistate versus two-state model. *Chem. Phys. Lett.*, **2006**. 439, 162–165.

[124] T. Kubař and M. Elstner. What Governs the Charge Transfer in DNA? The Role of DNA Conformation and Environment. *J. Phys. Chem. B*, **2008**. 112, 29, 8788–8798.

[125] A. A. Voityuk, K. Siriwong, and N. Rösch. Environmental Fluctuations Facilitate Electron-Hole Transfer from Guanine to Adenine in DNA π Stacks. *Angew. Chem. Int. Ed.*, **2004**. 43, 5, 624–627.

[126] X.-J. Lu, M. E. Hassan, and C. Hunter. Structure and conformation of helical nucleic acids: analysis program (SCHNAaP). *J. Mol. Biol.*, **1997**. 273, 3, 668–680.

[127] M. Büttiker. Absence of Backscattering in the Quantum Hall-Effect in Multiprobe Conductors. *Phys. Rev. B*, **1988**. 38, 14, 9375–9389.

[128] B. Popescu, P. B. Woiczikowski, M. Elstner, and U. Kleinekathöfer. Time-dependent view of sequential transport through molecules with rapidly fluctuating bridges. *Phys. Rev. Lett.*, **2012**. 109, 17, 176802.

[129] P. J. Acklam. An algorithm for computing the in verse normal cumulative distribution function, **2010**. Http://home.online.no/~pjacklam/notes/invnorm/.

1. M. Wolter, M. Elstner and T. Kubař, "*On the Structure and Stretching of Micro-hydrated DNA*", Journal of Physical Chemistry A, **115** (41), pp 11238 – 11247 (2011).

2. M.Wolter, P. B. Woiczikowski, M. Elstner and T. Kubař, "*Response of the electric conductivity of double-stranded DNA on moderate mechanical stretching stresses*", Physical Review B, **85**, 075101 (2012)

3. M. Wolter, M. Elstner, and T. Kubař, "*Charge transport in desolvated DNA*", J. Chem. Phys., **139**, 125102 (2013).

Last but not least...

An dieser Stelle möchte ich mich herzlich bei all denen bedanken, die zum Gelingen dieser Arbeit beigetragen haben.

Als erstes bei Herrn Prof. Dr. Marcus Elstner für die Möglichkeit der Promotion in einem interessanten und hochaktuellen Forschungsgebiet. Damit auch für die hervorragende Betreuung und Unterstützung in allen Projekten und Lebenslagen. Ohne Ihn und seine Arbeitsgruppe wäre die zweite Hälfte meines Chemie Studiums nur halb so motivierend gewesen und es wäre wohl nie zu dieser Arbeit gekommen.

An zweiter Stelle kann mein Dank nur an eine Person gehen: Dr. Tomáš Kubař. Beginnend mit meiner Masterarbeit hat er mich durchgehend mit allen Mitteln unterstützt, die ich benötigte um die Herausforderungen der alltäglichen wissenschaftlichen Arbeit zu bewältigen. Die in dieser Arbeit verwendeten Methoden zur Untersuchung des Ladungstransfers sind zu großen Teilen seinem Elan und seiner Hingebung zur Wissenschaft zu verdanken. Nicht zuletzt dafür, dass er unermüdlich unseren Rechencluster erweitert und am Laufen hält, mit besonderer Vorliebe auch Sonntags und mitten in der Nacht.

Natürlich möchte ich der gesamten Arbeitsgruppe Elstner danken, egal ob noch aus Braunschweiger Zeiten oder die aktuelle Besetzung in Karlsruhe. Es war immer eine sehr nette Arbeitsatmosphäre die immer wieder motiviert hat. In alphabetischer Reihenfolge: Jan Frähmcke, Michael Gaus, Alexander Heck, Steve Kaminski, Tomáš Kubař, Maximilian Kubillus, Gesa Lüdemann, Marco Marazzi, Sabine Reißer, Thomas Steinbrecher, Hiroshi Watanabe, Kai Welke, Benjamin Woiczikowski, Tino Wolter, und nicht zuletzt natürlich Sabine Holthoff, die jede Angelegenheit, sei sie auch noch so bürokratisch und/oder dringend gewesen, irgendwie geregelt bekommen hat.

Ausserdem danke ich den Korrekturlesern, die mir im letzten Abschnitt des Verfassens dieser Arbeit eine sehr große Hilfe waren: Tomáš Kubař, Alexander Heck, Gesa Lüdemann und Jan-Philipp Buch.

Schließlich danke ich meinen Freunden und meiner Familie für ihre Unterstützung!